What's

T0152623

I wrote this book for you. I wrote it for people *like* you and me who, in the face of the ever-increasing demands on our attention, are looking to create space between stimulus and response.

What *you* do next can be as big or as small as you'd like. If you're a small business owner, a solopreneur, or a CEO of a large company, what *next* looks like will be different. If you work in an office and face constant interruptions, or work from home and your biggest distraction is your own brain, what *next* looks like has to make sense for you, your needs, and your goals.

Maybe you'll pick one part of your day on which to focus. Maybe you'll create a handful of ActionStacks to make your days more efficient and effective. Perhaps you'll choose one specific goal to *Put Success in Your Way,* or maybe you'll do something helpful for *Tomorrow Guy.*

One of the easiest next steps is to sign up for my weekly newsletter here: robhatch.com/attention-newsletter. Each week I share new ideas, information, and approaches to implementing the ideas and information I've discussed in this book. When you sign-up, I'll send you a free pdf copy of my daily sheet along with a short video describing how I use it to *decide before I have to* and *Put Success in My Way* each day. If your next step is to explore what one-on-one coaching looks like, you can sign up for a free 30-minute coaching call at robhatch.com/coaching. I'd love to talk with you.

If you'd like to hire me to work with your team, or speak at your next event, visit robhatch.com/speaking to learn more.

More than anything, I'd love to hear from you. You can always drop me a line with any questions you might have at rob@ robhatch.com.

Thanks. I look forward to hearing from you.

Endorsements

Rob Hatch has created something new in a crowded field: a flexible method that reframes how we can do our best work better. Because he built these frameworks for himself and his own struggles, they're simple and practical. Because he's been teaching them to hundreds of people, you know they work. I use these methods myself and with my team. It gives us the flexibility to be freely creative in our creative work and solidly productive in our structured work.

Becky McCray, beckymccray.com

With *Attention!*, Rob Hatch offers an inspiring and practical guide that can support us all with putting greater success in our way. So many of us currently find ourselves living in a constant state of information overload and distraction that hinders our ability to focus on the things that matter most. The cutting-edge tools and frameworks included in this book help transform anxiety, overwhelm, and decision fatigue into new empowered habits that can lead to confident intentional action towards more meaningful success. This book should be considered essential reading if you are seeking to more effectively leverage your time, energy, and effort for the most impactful outcomes in these rapidly changing times.

Aime Miyamoto, Business Alignment Consultant

Rob has been telling me these ideas for more than a decade. As he's learned the concepts and tested them, he's taught them. More significantly, he's lived them. I've eaten his breakfast sandwiches when visiting his family. I've listened to the conversations with his family and seen the growth across time. I've watched his example and been shaped. I'm grateful that this approach to attentiveness is now available to people beyond Rob's friends and clients. I can testify that it's grounded in a deep caring for the people around him.

Jon Swanson PhD, Hospital chaplain and writer

Attention! is a spiritual call to intentionality and a neurological commitment to a practical minimalism. From opening premise to page layout, Rob Hatch's examples of personal pain points stoke the embers of curiosity and self-examination while each page turned promises to bring the reader to the Zen-like conclusion that less is more.

Robbie Grayson, Traitmarker

ATTENTION!

THE POWER OF SIMPLE DECISIONS IN A DISTRACTED WORLD

ROB HATCH

First published in Great Britain by Practical Inspiration Publishing, 2020

© Rob Hatch, 2020

The moral rights of the author have been asserted

ISBN   978-1-78860-145-0   (print)
       978-1-78860-144-3   (epub)
       978-1-78860-143-6   (mobi)

Every effort has been made to trace copyright holders and to obtain their permission for the use of copyright material. The publisher apologizes for any errors or omissions and would be grateful if notified of any corrections that should be incorporated in future reprints or editions of this book.

 Practical Inspiration
PUBLISHING

For Megin. You are the driving force for everything I do. Thank you for the time and space to make *Attention!* a reality.

For my Mom. You are a gift of kindness and strength. Thank you for providing a platform from which I could step out into the world.

Contents

About the author

Rob Hatch is the co-founder and president of Owner Media Group, providing strategies and skills for the modern business.

He is also a highly sought-after business coach and advisor serving executives and owners of organizations large and small.

As a speaker, trainer, and coach, Rob works primarily with business leaders and teams, guiding them through transitional moments in their organization.

Most importantly, Rob is dedicated to helping you do your work, better. This is accomplished through a private group coaching experience, through courses, and through business advisory and coaching offerings.

Rob lives in Maine with his award-winning photographer wife, Megin. They have four children and spend most of their time encouraging and supporting them as they pursue their passions.

Foreword
by Robert Brooks, PhD

I was trained as a clinical psychologist more than 40 years ago and during my career I have witnessed several major changes of focus in my profession. One of the most dramatic has been a shift from a so-called 'medical model' with its emphasis on 'fixing deficits' in people to a strength-based perspective that places the spotlight on identifying and utilizing each person's strengths or what I have referred to as their 'islands of competence.'

This shift is associated with the emergence of the field of 'positive psychology' and an increased interest in a concept that has been a major focus of my work, namely, *resilience.* Studies of resilience in both children and adults have attempted to identify individual factors as well as environmental forces that contribute to our ability to deal more effectively with stress and to thrive in the face of adversity.

In my roles as a clinician, consultant, and parent, I have often considered the question, 'How do resilient children or adults see the world and themselves differently from those who are not resilient?' This represented more than an academic question for me. I assumed that the more precisely we could identify those characteristics that defined the mindset and behaviors of resilient individuals, the more we could develop strategies to nurture resilience in both our children and ourselves.

I discovered that one key attribute of resilient individuals was their adherence to a lifestyle rooted in what I have labeled 'personal control.' Resilient people demonstrated an impressive capacity to focus their time and energy on situations over which they had some control or influence rather than on situations in which they had little, if any, impact.

Personal control represents a very significant quality in determining our emotional and physical well-being. While conducting therapy I constantly witnessed the negative outcomes associated with the absence of a sense of personal control. I heard accounts of people who based their happiness on others changing first (e.g., 'I would be happy if only my wife treated me better'), or who continued to cast blame on a particular situation for the misery they experienced in their lives (e.g., 'Why did I have to be born with a learning disability? It's not fair!'), or who felt so pessimistic that they could not even consider steps they might initiate to improve their lives. Often a 'victim' mentality dominated their existence.

In contrast, through interviews and correspondence, I heard from individuals who when unhappy would examine what they could do differently to improve the situation, of children or adults with learning problems who moved away from a position of asking 'why me?' to adopting the attitude, 'I had no control over having learning problems; what I do have control over is finding the best ways to learn, and the best people who can help me.'

Given these first-hand accounts I gathered from resilient people, I voiced the opinion that 'we are the authors of our own lives, that while events occur beyond our control, what we have more control over than we may realize is our attitude and response to these events.'

Yet, even while subscribing to this viewpoint, I am fully aware that we live in a world in which an increasing number of people feel overwhelmed on a daily basis. They believe they have little, if any, control over what transpires in their lives and that the time needed to meet all of their responsibilities is in short supply. Distractions are everywhere – just observe the number of children and adults who throughout the day are unable to free themselves from their mobile devices or computers.

Live conversations have taken a backseat to brief texts and constant texting often shifts our focus from the important task at hand. Hours are lost rather than gained in our attempt to multi-task. In the middle of completing a project, we are drawn to check and answer our texts and emails, or even begin a new project before the current one is completed.

Eventually most of us realize that our time and energy are being squandered and any sense of personal control is lost. As one of my patients lamented, 'I'm spending more and more hours accomplishing less and less. I seem to be spending less time with my family and I'm less productive in my work.'

He added, 'I just don't know where to begin to change things. I try to cut back on some things, but then I just add more hours doing other things that are also not productive.'

Early in my career I was taught that if individuals knew they had to make changes to improve their lives but failed to do so, they were being 'resistant' to implementing new behaviors, that deeply rooted, unconscious forces were working against them confronting their problems. In some instances, this might certainly be the case. However, what I was to learn that seems so apparent now, was that more often than not the main obstacle to taking control of one's life was that people lacked a plan, a system, or well-thought out strategies to move forward. They felt trapped with no compass to guide them.

In the absence of a clear blueprint for action, attempts to improve one's life are likely to eventuate in unsuccessful outcomes. Such circumstances may invite a vicious cycle. Negative outcomes intensify feelings of pessimism, which result in people giving up after just a brief time, convinced that they cannot make changes in their lifestyle. Sadly, they resort to the same counterproductive script they have followed for years – as unsatisfactory as this script has been, it is the only one they know.

It is important to note that roadmaps do exist that can guide us to realize greater personal control, flexible structure, and happiness in all arenas of our lives. One is Rob Hatch's very impressive book *Attention!*. Rob has done a masterful job of not only detailing the problems we face living in a world full of distractions – distractions that drain our time, attention, and energy from greater satisfaction and achievement – but, very importantly, he provides very specific, practical, realistic techniques for bringing order, purpose, and success to our lives.

Rob has a wonderful ability to introduce us to concepts that are understandable and can be translated into action. The material is further enriched by the many personal experiences Rob shares and his accounts of what he learned from both the positive and negative events in his life.

This personal quality has contributed to the creation of a very reader-friendly, informative book that on each page conveys a sense of empathy and a recognition that we have all struggled with obstacles in our lives and that there are strategies we can use to confront the challenges we face.

An example of one principle that Rob proposes is *Put Success in Your Way*. In explaining this principle, Rob identifies three core elements, that I am certain will prompt much self-reflection. They include:

1. Willpower is a limited resource.

2. Decisions are distractions.

3. Habits are a powerful force to which we are biologically prone.

The clarity with which Rob elaborates on these three elements and his recommended realistic strategies to *Put Success in Your Way* will be welcomed by all readers. They are seemingly simple ideas, which if followed can result in significant lifelong positive changes.

Earlier I highlighted a key conviction that guides not only my professional activities but also the actions I assume in my personal life: 'We are the authors of our own lives.' Thus, I was delighted to read another of Rob's principles that resonates with being authors of our own lives, namely, 'You are the architect of your own system.'

In discussing the word 'architect' Rob emphasizes that 'we are in charge of building the experience we want.' Via the words 'of your system' Rob encourages us to examine the current system we use, what parts of that system require modification, and what steps we can take to make these modifications.

The message housed in all of Rob's principles and strategies is that we are capable of discovering ways to remove counterproductive scripts that burden our lives. Rob recognizes that re-writing certain scripts may, at times, be experienced as a formidable task and that initial efforts may not prove effective. However, his recommendations for learning from both our successes and setbacks can provide the confidence and direction we require to lead a life of our choosing, one in which we feel empowered, and one in which we truly believe that we are leading a life in concert with our values.

I am certain that Rob's book will become an invaluable resource to be read and re-read as we search for a more purposeful, meaningful life.

Introduction

Do you value your time and attention as much as marketers do?

It's no secret that marketers have been vying for our attention for hundreds of years. But the stakes have risen and the capabilities available to them are beyond what many of us imagined.

Tony Fedell, the founder of Nest, was also on the team that created the first iPhone. He admittedly has some regrets about the unintended consequences of his creation.

'I wake up in cold sweats every so often thinking, what did we bring to the world?' he says. 'Did we really bring a nuclear bomb with information that can – like we see with fake news – blow up people's brains and reprogram them? Or did we bring light to people who never had information, who can now be empowered?'[1]

The answer, of course, is both. There's no doubt the iPhone has been transformational in its importance. The access to information and knowledge with this technology has empowered and democratized entire populations of individuals.

---

[1] Schwab, K. (2017, July 7) 'Nest founder: "I wake up in cold sweats thinking, what did we bring to the world?"' *Fast Company*, www. fastcompany.com/90132364/nest-founder-i-wake-up-in-cold-sweats-thinking-what-did-we-bring-to-the-world

But spend an evening with a family of four. Listen to parents who openly wrestle with the impact it has on their children, as they struggle to set limits. There are full-blown arguments between parents and their children driving a wedge in family relationships.

The irony, of course, is that while parents are trying to figure this out, they are simultaneously interrupting these important conversations with a quick check of their own phones.

We weren't equipped for this. We weren't prepared for just how quickly these powerful tools would capture our attention and, more importantly, our time.

It's not *all* about the technology

Let me be very clear. I enjoy technology. I enjoy my iPhone. I've written much of this book on my Mac or MacBook, and even, at times, my phone. I used Google docs and other apps.

My children all got their first phones in middle school. That was our rule. And, yes, my wife and I have spent hours talking about how to best handle conversations about how and when they are used.

Personally, I land on the side of believing that, for all its shortcomings, for all the things that kept Mr. Fadell awake at night, I am grateful for all the transformative power he helped bring to the world.

Distractions are everywhere. This is only one example.

This book wasn't written to tell you to take a digital fast. That's up to you to decide. And that's the point.

The flow of information and noise coming at us is overwhelming.

But we get to choose what we let in and how we direct our response to it.

I think we've lost sight of that a bit. We've welcomed that endless stream of information into our heads. More than that, we actively seek it out. And in doing so, we've lost the space in between.

I believe in our ability to reclaim some of what we've lost. I believe in our ability to choose where we give our attention, for what purpose, and to whom we give it.

There is a quote that has been mistakenly attributed to psychiatrist and Holocaust survivor, Victor Frankl: 'Between Stimulus and Response, there is a space. In that space is our power to choose our response. In our response lies our growth and our freedom.'[2]

Our freedom to choose is perhaps the highest form of wealth and power.

When we are able to consciously direct our time and attention to things that matter to us, we are able to transform our lives. We are able to affect the trajectory of our careers, start a business, build connections, and deepen our relationships with friends and family.

---

[2] Unattributable, https://quoteinvestigator.com/2018/02/18/response/

In a time and culture increasingly burdened with anxiety and stress, the freedom to choose provides respite. These conscious acts of choosing how we focus our attention are how we create space to what truly matters to us.

This is where the power of simple decisions begins.

It's in the narrow spaces between what we see and what we do.

The more we seek that space, however small, the greater our ability to widen it, reclaim our attention, and live a life of intention.

PART ONE

The state
of things

CHAPTER 1

●

The problem:
our distracted world

It's 5:30 a.m. and your alarm goes off.

Of course, it's not an alarm clock. It's a ringtone you carefully selected and scheduled on your phone.

Maybe you hit snooze, but more than likely you turn it off and immediately unlock your phone to check something, though you're not sure what yet.

It could be the score of the game you fell asleep watching the night before, but more than likely you aren't looking for anything in particular, you're just checking.

You open Instagram or Facebook to see what's there, scrolling past several posts and then flip over to email because you just remembered you were waiting on an email from a client.

As you start to scroll through your inbox something else catches your eye. It's an email from your boss asking a question about the presentation you've been working on. It's an easy question to answer, so you sit up in your bed, grab your glasses, and shoot off a quick response.

You go back to your inbox, trying to remember what you were looking for in the first place, and you see that your boss replied almost immediately. There's a twinge of guilt that she's already up and working. *Of course, she's probably emailing from bed as well. It's what we do.*

You read her response and she asks if you can meet today. You quickly check your calendar and see that you can. While you're there, you notice you forgot you had a phone call with a new client later.

You quickly switch back to your email to search for your last communication with the client to 'refresh your memory' about the meeting. Suddenly, a slight panic hits you because you forgot to respond to your boss about the meeting. After shooting her a quick reply that you *can* meet, you remember that you originally opened your email to look for something else.

Eventually, you find what you were looking for and give it a quick scan. You don't need to reply, but decide to shoot a quick, 'Thanks for this. I'll look it over and get back to you later.'

Your alarm went off at 5:30 a.m. You've been awake now for seven minutes.

You go back to your boss's email to confirm the meeting time, and schedule it in your calendar. When that's done, you breathe a sigh of accomplishment and pop back over to check Facebook again. After all, you need a break from all the work you've done.

You scroll for a few minutes. A friend shared an article that looks interesting and you start to read it. Halfway through that you notice the time. Now you're behind schedule.

A quick shower leads to looking for the pants you were hoping to wear and wondering if you know where your favorite shoes are.

You ultimately decide on another outfit entirely but not before leaving your closet a mess and your dresser half open. Vowing to deal with it later, you realize you're even further behind, and start looking for a quick breakfast as you make the coffee, let out the dog, and feed her.

In the rush, you start feeling anxious about the day and decide to check your phone again to see if anything else has come in that you need to be 'ready for.'

Can I stop now?

I know this sounds familiar because some version of this plays out each morning in the homes of nearly every working adult I know.

What follows is a deliberate process to understand and address the forces at play in our daily lives that contribute to the noise and distract us from the life and work to which we aspire.

Some of this noise is external. But often the loudest is internal.

As we progress, we will identify opportunities to leverage these forces for our own purposes. We will use the power of simple decisions to reclaim the space between stimulus and response and direct our attention to what matters most.

At the end of each chapter, I'll leave you with a few thoughts to consider.

My hope is they will help you recognize the challenges, but also frame up an approach that works within the context of your life to address them.

Red-dot reactions: the stimulus

My business partner, Chris Brogan, has a saying: 'Email is the perfect delivery method for someone else's agenda.'

But it's not just email. It's *every* notification we receive on our computers, tablets, and phones.

The dings and buzzes and banners on our devices have us on a very short leash. The default setting for most of the software we use for communication is to let us know when something new occurs. In our work, we've come to accept that we need to be responsive and available, and so we allow a constant stream of interruptions to grab our attention.

It's gotten to the point that even when nothing happens, we instinctively check our screens for the *little red dot* telling us someone, somewhere has done something.

Our red-dot reactions have led us to live our lives on constant alert. Our reactions are so swift, so instinctive, we are leaving virtually no space between stimulus and response.

If you've spent any time in a crowded restaurant and someone in a nearby table gets a text message, you've seen how every person within earshot picks up their phone to check to see if it's theirs.

I've done it myself even when I know the sound I heard is not the same as mine.

The irony, of course, is that in all of our rushing around, we are genuinely trying to find that space. We chase it every day, but it never materializes.

We've accepted the default settings. We've given permission to everyone we know to interrupt us at any moment.

We don't *allow* ourselves the time to define what we want our lives to look like.

We don't *choose* how and when we want information delivered or which notifications we want to get through and those we should filter out.

We don't *recognize* the power we have to direct our time and attention to things that matter.

Of course, we like to blame technology for this. It certainly bears its share of responsibility, but even with the demands of technology, we can find the space for choice.

And, perhaps, set up technology to serve us.

False choices

We are inundated with a variety of choices each day.

Marketers provide these choices under the guise that we are individualizing our lives. They promote the idea that these choices enable us to create a story of who we are. The brands we choose are an outward statement of the tribe to which we belong.

But there is a presumption of conscious action on our part. The choices, options, and features create an illusion of control.

But the sheer volume of options, from which coffee to which flavor of pasta sauce in the grocery store alone, can be downright debilitating.

The irony of having so many options is twofold.

The *first* is that with so many choices, we are paralyzed and don't make any.

The *second* and more common is that with so many choices, we choose quick and convenient over thoughtful and relevant.

Simply put, more choices result in our choosing poorly.

This isn't just because we are prone to make bad choices. It's not even the fault of the marketer who is attempting to take advantage of our time-crunched, overwhelmed existence. Though there are kernels of truth in both.

It's because we haven't defined what constitutes a good choice for us before we're faced with the prospect of deciding.

And while, yes, we have come to accept the idea that our purchases are indeed a means for defining our sense of ourselves as individuals, we have determined ahead of time that we are the type of person who wears a particular brand, for example.

But when it comes down to the temptation of shoes on sale versus the balance on our credit card, we err on the side of spending, not saving.

This isn't a rant against rampant materialism. Trust me when I tell you, I love new shoes.

Rather, it's a call to flip the process.

It's a call to define yourself and your values.

It's a call to ground yourself in the life you most want to create as a filter for making decisions that align with your vision and values.

Time and money

We bet against our own interests. We claim we have no time, yet we are more than happy to binge entire seasons of *The Marvelous Mrs. Maisel.*

Our spending habits are also not in alignment with our long-term goals *or*, as is often the case, our short-term realities.

Nearly 78% of Americans have little to no savings and live paycheck to paycheck.[3] They carry significant amounts of debt, and have no ability to weather an emergency.

Those realities are a recipe for stress and anxiety. Yet, we continually make decisions that perpetuate the cycle. We take on too much debt in the form of newer and nicer cars, boats, houses, and, yes, even our phones.

---

[3] CareerBuilder (2017) 'Living paycheck to paycheck is a way of life for majority of U.S. workers, according to new CareerBuilder survey,' http://press.careerbuilder.com/2017-08-24-Living-Paycheck-to-Paycheck-is-a-Way-of-Life-for-Majority-of-U-S-Workers-According-to-New-CareerBuilder-Survey

I recently had the same conversation with two people. One was my 16-year-old son; the other, a successful CEO.

Neither one understood the process of upgrading a phone. Specifically, that when your mobile carrier tells you that you're 'eligible for an upgrade,' what they are actually telling you is that you are eligible to apply for a $1000 loan. They, in turn, will conveniently spread the payments out over two years at $50/month so that you can have the newest phone.

This is what happens when our decisions are not aligned with our goals.

This is what happens when we purchase as a means of defining who we are.

This is the value of your attention.

More importantly, it is the value of your distraction and overwhelm.

When companies make it easy for you, it's probably a good idea to take a breath, find space, and then decide.

And just in case you're thinking this is some anti-consumerist rant, I chose to upgrade my phone recently.

They did make it easy, but I took a breath, just to be sure.

Everything in our way

For many of us, the demands and chaos of daily life leave little room for feeling as though we are in control. Our lives don't seem to have the rhythm they once did. This may be why we 'long for yesterday' (thanks Paul McCartney).

There have always been times when life seems to have a cadence we can count on. That cadence gives us a sense of knowing what to expect and helps carry us through our days.

As with anything, sometimes this cadence gets out of sync. The beat doesn't skip, but the sound of a hundred other instruments becomes audible and the clear rhythm has been swallowed by a cacophony of demands and the volume of interruptions.

But it's still there.

It shows up in seasons. Not necessarily the seasons of nature or even those measured by holidays, but we can identify times in our lives when things just work well.

There are many reasons for this. Sometimes it seems entirely circumstantial. We may think it is the result of the stars aligning a certain way.

As much as we are aware of the effect of circumstances and forces at work around us, it's important to look for the clues to our success and what role we have in choosing to set the rhythm.

Our personal and work lives have become intertwined. Constant connectivity has left us with little space for the quiet movement from one thing to another. Boundary lines have blurred such that every aspect of our lives feels cheated and the idea of knowing what to do next feels impossible to truly grasp.

I am convinced that the amount of information we consume each day is shortening our life. It may be a slow death, but the articles we click to 'stay informed' or for entertainment are sucking up precious hours with not much to show for it.

Add up the time you've spent reading various commentary over the past six months. Is it an hour a day? Is it two hours? More?

It's not that watching your favorite shows or reading articles your friends shared on social media is inherently bad, if that's what you've decided to do.

But we aren't deciding, we are reacting and justifying.

Several years ago I was approached by a client named Franco.

Franco was, by all accounts, a successful salesperson. He provided a very good life for his family. He had been recognized at times as a top sales performer in his company. From the outside, life looked pretty good.

However, Franco suffered from doubt, anxiety, and frustration. At the time, he possessed a solid procrastination game. He's also incredibly smart and charismatic. You get the immediate sense that he's someone you would trust and would give your business to very quickly.

Franco was also a voracious reader. Most successful people are. He devoured books and blog posts from the top marketers and sales gurus, always on the lookout for a new idea or tip.

As we began our work together, I was quickly taken in by his charm and admittedly thrown off a bit by his outward success and never-ending curiosity. Our initial conversations were lively. He peppered me with questions, seeking my opinion on the latest morsel of advice he had just read.

But as we continued, it became apparent that his incessant pursuit of the next idea was a huge part of what was holding him back.

His actions were inconsistent. His days and weeks were up and down in terms of productivity.

But smart, charming people with a knack for hustle always seem to find a way to pull things together in the end. Until of course they don't.

Eventually, the cycle of feast and famine wears them out. These ups and downs, particularly when your income is derived from business you generate, creates a din of white noise in the form of anxiety. What starts as a soft hum, turns into the debilitating reverb of procrastination.

One of the challenges was that Franco's curiosity had no purpose or grounding point.

He had convinced himself that reading was the equivalent of work.

His desire to devour every tip and trick he could find on the subject of sales and marketing felt like research. But all that research wasn't being applied consistently, and there was no way to measure the results.

Because of his propensity for procrastination, he was simply in search of a quick fix.

What was truly unfortunate was that he didn't trust himself. He had never taken the time to embrace his personal skillset. Instead, he counted on being able to pull things off, but never found a way to leverage his skills consistently.

During one of our conversations, I told him he was no longer allowed to read, anything.

I know!

What a weird thing to suggest.

But in this case, reading was an excuse to avoid execution. It became an escape. He justified it as learning or researching. He felt like he was putting in his '10,000 hours.'

What he was actually doing, was reading about people who *had* done the work.

The concept of 10,000 hours introduced by Malcolm Gladwell[4] isn't just about a number, nor is it about reading to acquire information.

It's about being deliberate in your practice of a skill in pursuit of mastery. And as important as the practice of skills are, the step of synthesizing what you have learned and the integration of that into action is as important.

Franco's constant pursuit of information wasn't only affecting his professional life. The pursuit of distraction was keeping him from engaging with his family and enjoying his life.

---

[4] Gladwell, M. (2011) *Outliers: The Story of Success*, New York: Back Bay Books; reprint edition

Of course, not reading isn't going to solve the problem. The same is true with a digital fast. But we had to take control at the point of delivery. Stopping the flow of information is only temporary. It's what comes next that matters.

My question to Franco would be the same question I might ask you if we were working together: *What does that look like?*

Specifically, *what does it look like when things are going well?*

This is about *his* seasons.

In Franco's case, I asked him to write out his sales process for when he lands his best clients.

I wanted him to lay out the method by which, if he followed it step-by-step, he would almost certainly close the sale. Interestingly, he had never considered that he might have his own repeatable method for selling.

Here's another question to get used to: *And then what happens?*

As he began to describe each step, he had to ask himself, *and then what happens?*

We reviewed and refined his process together.

We combined steps to make things more efficient.

His assignment then on was to focus only on executing *his* methods consistently and *not* to look outside for a new tip or trick. No more reading (for a while at least).

Several months later, Franco began receiving awards (and bonuses) for his performance. His boss noticed the change. His colleagues in his company reached out to him, seasoned salespeople as well as relative newcomers sought his advice and coaching in his methods.

He was invited to present at his company's international conference and eventually started a new business coaching other salespeople on his methods.

With his boundless curiosity, I couldn't keep him from reading and devouring loads of information on sales and marketing forever.

The difference now is the information and ideas he seeks out have a purpose. He has a core understanding of his mission, values, and ownership of his methods against which he can synthesize new ideas with his method.

Establishing a sense of who we are and what makes us successful is far more powerful than seeking new tips and tricks to fill the gaps created by our failings.

Red-dot reactions: the attention vortex

While technology is not entirely to blame, there is no doubt that we've not encountered a phenomenon with the same gravitational pull as the mighty red dot.

Software companies employ myriad psychological tactics to compel our attention and hold it. And with good reason.

Attention, time spent in an app, *directly* translates to profit for these companies. Which begs the question, *if these companies have assigned a value to your attention, do you value it as much as they do?*

Of course, the same was always true of television and newspapers before that. Joseph Pulitzer was the master of the headline. He understood the concept of 'clickbait' long ago.

Today the delivery methods are much more sophisticated. The speed at which we are able to access and consume information has changed the game entirely.

We grab our phones the moment we receive a notification to see who liked our post or left a comment. And when we open the app, you can be sure we'll be there awhile caught in the attention vortex, even if we were in the middle of doing something else entirely. Maybe even writing a book.

We've all done it. We're not even sure why we opened it in the first place and suddenly we're scrolling and clicking.

Every click takes us deeper. Everything we read takes more time.

Unfortunately, it doesn't stop there. Even when we close the app, we're left with lingering thoughts that we bring into our conversations with friends and family. And while time with family sounds wonderful, how much time is spent in conversations that begin with the phrase, *'Did you see that thing…,' 'Can you believe…,'* or *'Did you hear…'?*

More often than not, those conversations are not about connecting in a meaningful way with our friends or family.

They are not rooted in growing as a person or moving our business goals forward.

It's not that they aren't fun or entertaining, but they are not terribly useful.

And let's be honest, those conversations aren't about the substance of the article and your point of view on the ideas. We're reacting to the headline.

The math

In recent years, companies have begun to provide personalized 'Screen Time' reports on the time we spend on our phones. I receive a weekly summary of how I spent my time and which apps got most of my attention.

It's shocking how easily two hours passes when we're scrolling.

Of course, our average screen-time consumption is much higher, but here's some math to consider:

182 days of spending two hours a day, adds up to 364 hours. That's a full 15 days in a six-month period.

We spend one month a year consuming information that, in all likelihood, doesn't help *you* or *your business* in a meaningful way.

More math

In one year, we are spending 728 hours clicking on headlines or scanning them as we scroll by.

We work between 8 and 10 hours a day (give or take).

728 hours / 8 = 91 work days a year.

Most people work 5 days per week.

91 days / 5 = 18.2 weeks a year.

That's over 35% of our work year we are giving away. And did I mention that two hours a day is well below average?

'Well that's depressing, Rob.'

I'm sorry about that, I really am.

And I know there are legitimate reasons for staying informed, or doing research. And entertainment is a great reason to use technology.

But if you're anything like me, the lines get blurry.

We lose track of time easily. We aren't aware of how much time we're letting slip away.

We aren't deliberately seeking information to grow or help us solve a problem. We are consuming randomly. We take in what's fed to us.

It is called a 'feed' after all.

But it doesn't have to be this way.

Reclaim the space

These tools and platforms can be customized to our liking.

And it's our job to make sure it's filled with the information and resources that serves us.

I have conversations with lots of folks who are feeling unfocused. These are smart people with good, healthy businesses.

They are experiencing information fatigue. Our minds tire from the overload. We're not allowing our brains the time and space for sorting and synthesizing of information they take in each day.

Look at the numbers

When you don't know how you are *spending* your resources, it's hard to make changes. As depressing as it might be, one way to gain control of a situation is to take a good look at the numbers.

A lack of focus is a symptom. It's a sign that something else is at play. Sometimes there are legitimate distractions. *An illness in the family takes a tremendous toll, for example.*

More often than not, however, we are allowing the noise to shape our time. We are accepting the distractions as a 'way of the world' and we don't have to accept that.

It starts by getting a clear picture of how we are spending our time and resources.

Move past the guilt

Closer looks like this are scary.

We know it may not look pretty. The temptation will be to start blaming ourselves.

Cue Robin Williams in *Good Will Hunting*: 'It's not your fault.'

Well, it may be a little bit your fault, but that's not exactly helpful now, is it?

I have always had a knack for carrying guilt around like a yoke. I am keenly aware of how I 'got myself into this.' But it doesn't do me much good to dwell there. Quite frankly, it's a waste of even more time.

However, facing up to where you are is the best way to figure out a path to where you want to be.

We can't wallow in guilt. We have to move forward. And if you want things to be different, consuming more information isn't likely to be the solution.

Of course, it's not that easy, until it is.

Take a moment to let this settle in.

Did you identify with that feeling of overwhelm each morning?

Did you see yourself in Franco's story?

Did the red-dot reactions resonate that you actually picked up your phone to see how many there are?

It did for me. Sometimes it still does.

I want you to know that I share this challenge with you. I am not immune. I do not have a natural inclination to organization and productivity. The ideas, methods, and systems I'm sharing with you are what I use every day to help tame the demons of my own distractions.

What does that look like?

What would it look like if you were able to cut out half of the interruptions that make their way into your brain?

Where are they coming from?

Spend the next 24–48 hours just recognizing and noticing what you allow in.

CHAPTER 2

●

This isn't working: we weren't prepared for this

I'm a fan of building on strengths.

I'm a fan of taking what you know has worked and turning it into a repeatable system.

I'm not talking about a prescription or an exacting and rigid formula. Systems should be flexible. They should grow with you and adapt to your changing circumstances.

A great place to start with any system is to look at what has worked… for you.

'You'll never make it!'

This phrase would never make the cut for an inspirational poster.

I swear though, there are days when the loudest voice in my head is holding up this sign.

What's worse is that this voice speaking to me is well-intentioned. It wants to save me from yet another false start or failed finish. That's helpful, right?

If there's one thing that continually rears its head when I'm attempting to start something new, it's that voice reminding me of the long list of half-finished projects, unmet goals, and unrealized ideas.

I never made it.

But I thought we learn from failure?

We should.

But, honestly, we don't make time to truly consider what went wrong and learn from our mistakes.

It's not as simple as when we first learned what to do and what not to do:

> Touches hot stove.

> Gets burned.

> Knows not to touch the hot stove again.

And don't get me started on sticking your tongue on a frozen metal pole.

No, failure is never my go to when it comes to learning and developing a system.

I'm not looking for something to avoid. I want something upon which I can build.

This isn't working

The cult of *learning through failure* is failing us.

This idea that we learn by failing is misunderstood and misapplied. In fact, many of the quotes from famous and successful people diminish the word.

Thomas Edison is probably the most quoted in this area:

> I have not failed. I've just found 10,000 ways that won't work.[5]

Read that sentence closely.

Edison rejects using the word to describe anything he's done. He's pointing us instead to the 10,000 attempts and efforts.

That is the intention and purpose of the *learning from failure* philosophy.

It's meant to encourage us to press on and try again. And if we truly learn, our next attempt won't be from the same place as before.

---

[5] Attributed to Edison, T., https://quoteinvestigator.com/2012/07/31/edison-lot-results/

But in order to try again, one needs to get back to solid ground. Edison doesn't seem to indicate that failure is the place where he finds that.

Even the phrase 'back to the drawing board' assumes that you have a place to begin from. You have an idea, an inspiration, and in failing you have at least *one* attempt under your belt. And so, they are not failings at all, these are strengths.

An attempt is something new entirely, and the ground from which you leap, however small and shaky, is still something beneath you and a strength.

If failure does anything useful, it narrows our options. If we focus on what worked, even just a portion of it, if we attempt to replicate our success in some manner, and apply it in new ways, we are building on the success of previous attempts.

Learning from failure is more about eliminating bad options and applying our strengths in deliberate ways to accomplish something new.

Our first strength is the platform on which we stand.

The myth of 'grit' and the self-made person

We've bought into this notion of people being self-made.

We idolize those individuals as though everything they have accomplished is the result of personal grit in the face of challenges others simply weren't willing to persevere through.

There are certainly some very inspiring stories and examples. However, we miss something when we focus on the 'self-made' ideal.

A more holistic view would most certainly reveal innate gifts of talent, opportunities of circumstance, and a healthy dose of dumb luck that have contributed to each success story. But that's not as sexy.

I grew up in small town in Central Maine in a middle-class family although, in my first four years, we lived in a small mobile home. Middle class was something we would grow into.

When I was four years old, my parents bought a modest house in a quiet, friendly neighborhood.

Those two environmental circumstances alone are significant. My parents were intentional in their decision to raise a family in this town. Its schools were (and are still) well regarded. The neighborhood was also a deliberate decision. We lived a few houses away from people who helped raise my father.

Life was easy for me as a child. I was smart and did well in school. My parents encouraged me, signed me up for sports and scouts. They participated actively in our community, and showed up for everything I chose to join.

My mother and father had a firm but gentle approach to discipline. I always knew where the lines were, and like most children I found ways to push, for which I received my share of punishment.

But in *all* cases, I knew I had their support. I never once doubted their love, pride, care, and concern for me and my sister.

One of the clearest memories of their support happened in the fourth grade. We were attending a parent information night to learn about the school band. Much to my father's surprise, it was not just informational. Decisions were being made. Instruments were being distributed for which a 10% deposit was required.

At the time, it was $60, which works out to be the equivalent of $200 as I am writing this.

I was just 10 years old at the time, but I could see the pain on my Dad's face. I knew full well that paying $60 unexpectedly was a stretch for them.

I'm not sure how but, unexpected as it was, he put down the $60 for my $600 saxophone.

Of course, there were kids in my school whose parents didn't bring them to that meeting. I know there were kids whose parents didn't have the ability to lay down $60. I'm not sure mine did either.

I continued to play for four years. I wasn't great. I wasn't terrible either. That decision on that night in that room opened a door.

I learned to read music.

I learned how to practice.

I learned how parts of a band combine together and support each other.

I learned to listen and follow the lead of the conductor.

I performed in front of groups, competed in front of judges, and marched through our town in a parade.

I am not self-made.

Even if I practiced hard, in my room, by myself, and became a successful professional, nothing about the experience would have been 'self-made.' I started with an advantage. I had supportive parents, who had enough money to put down a deposit for the saxophone (even if it was a stretch), excellent teachers, and grew up in a wonderful town.

I am not self-made. That is a strength.

The advantage of platforms

You'll notice in the list of the things I learned that each one provided me with a platform, a foundation from which I can step into a whole host of other opportunities:

- Supportive parents who drove me to band night and paid a deposit for my instrument – *Platform*
- A school with a music department and qualified teachers – *Platform*
- Being taught how to read music – *Platform*
- Performing in front of others – *Platform*

While none of these guarantees success and not having them doesn't result in failure, the fact of the matter is I was given platform after platform on which I could stand and step from into whatever might be next.

Each one represents an achievement. Each one afforded me an opportunity to move up a level.

I didn't continue to play in high school. I chose to pursue other interests. Here's what didn't happen:

- No one told me to 'gut it out.'
- No one told me I was a quitter.
- I didn't feel as though I had failed as a musician.

My memories of it and the lessons gleaned are actually grounded in the successes of my experiences.

Learning to play an instrument is a great example of learning from success.

At its most basic, the notes sound better when you hit them. You can feel the difference.

When I played a wrong note, I wasn't focused on lessons from failure, I learned by taking what I knew (the notes and the fingering) and building on that.

Each new note was built upon the foundation of knowing how to hold the instrument, where to place my fingers, and trying something new and different.

Music isn't about learning from wrong notes and avoiding failure. It's about learning what works and building something that sounds beautiful.

There has been some debate about the limits of willpower. Do we have an inexhaustible supply available on demand? Do we run out? Can we replenish it?

Studies indicate that when exerted or tested, willpower has the potential to, in effect, become depleted.

In his famous 'cookie' study, Roy F. Baumeister et al. sought to understand the potential impact of taxing our willpower.[6]

To demonstrate this, Baumeister et al. asked separate groups of hungry college students to perform a task.

One of the groups was presented with a plate of warm cookies at the center of the table. They were told that they could eat them, but only when they had finished the test.

The other group was presented with a plate of radishes at the center of the table and given the same instruction.

The results demonstrated that the groups presented with the cookies did not perform as well on the task.

Baumeister et al.'s conclusion was, due to the exertion of willpower required to *not* eat the cookies, it taxed the students' willpower such that it negatively impacted their performance.

For the moment, let's accept Baumeister et al.'s conclusion that in the moment, it is possible to place a strain on our willpower reserves.

---

[6] Baumeister, R.F., Bratslavsky, E., Muraven, M., and Tice, D.M. (1998) 'Ego depletion: is the active self a limited resource?' in *Journal of Personality and Social Psychology*, 74 (5), 1252–1265, Case Western Reserve University, Cleveland, OH

That in order to perform well in a given situation, we would want to minimize the extraneous drains on our energy, to focus our willpower on the most important tasks in front of us.

So, the question becomes, why would we knowingly subject ourselves to this strain?

Why would we continue to allow obstacles, even seemingly insignificant ones such as the temptation of a cookie, when we can remove them and focus our willpower more effectively?

Remove and replace

Shawn Achor is an author, psychologist, and researcher in the field of positive psychology. While a professor at Harvard, his course was once the most popular among students. In his first book, *The Happiness Advantage*, Achor shared a personal story about his goal of learning to play the guitar.[7]

Like many of us who endeavor to acquire a new skill or develop a new habit, he had planned it all out. And as an expert in the field of psychology, he understands how human behavior works, which, of course, means he has a distinct advantage over many of us.

Like many of us, Achor's enthusiasm for this new effort led him to make a spreadsheet detailing when he would practice, for how long, and for how many weeks; it was beautiful.

---

[7] Achor, S. (2010) *The Happiness Advantage*, New York: Broadway Books

Also, like many of us with a new goal, he started out strong. Each night he would grab his guitar from his closet, practicing regularly in the early days. He was, by his own account, making progress and feeling pretty good about it.

A few days into the process, instead of grabbing his guitar, he found himself returning to old habits, sitting on the couch, and turning on the TV.

He succumbed to the forces that we all do. At the end of the day, we are tired and we just need to relax for a few moments. So, before we start anything new, we plop down in a comfy chair and grab the remote.

As an aside, let me say that there's nothing inherently wrong with relaxing at the end of the day by watching a bit of television. Unless, that is, you had plans to do something else.

As soon as Achor gave in to his old, familiar habits, it was the end of his plan.

He had probably spent as much time planning as he did in an earnest effort to implement it.

Sound familiar? It did to me.

I could spend days laying out a plan, especially a plan to change a habit or build a new one. We've all been there. You've been there.

The good news is that Shawn didn't let it end there.

He began to realize that the reason he stopped practicing was that it was all too easy to sit on the couch, pick up the remote, and turn on the television. And his guitar was nowhere in sight. In fact, it was in the closet.

Of course, it's not that the closet was on some remote island. It was a mere 10 or 20 seconds away. But at the end of a long day, his resources depleted, it felt like an eternity.

He then tried something so simple, it's almost laughable.

He removed the guitar from his closet and placed it on a stand, between the couch and the TV.

In addition to that, he removed the batteries from the remote and put those 20 seconds away, in the closet.

He didn't just remove the obstacle. From this point on, he was faced every day with the guitar and didn't have to motivate himself to get up and get it and sit down and practice. He didn't even have to think about it anymore.

He made the old habit harder and the new habit almost impossible not to follow through with.

This is an example of what I call, *put success in your way*. And it's the forces at work against Mr. Achor, against all of us, that make it both useful and necessary.

Put success in your way

The process of identifying and placing the most important elements to successfully complete a task directly in your path, including removing any obstacles or distractions beforehand.

Several years ago, I was the Executive Director of a non-profit organization in Western Maine. My office was a 40-minute commute each way. At the time, my wife was home with our (then) three very young children.

As with many leadership roles, especially in the non-profit world, you are required to wear many hats. There are many demands on your time, not the least of which are the employees you serve and support.

As a leader, it was my desire to be available to everyone. As such, I declared an 'open-door' policy. Upon reflection, this was a deeply flawed approach. There were roughly 40 employees at this location and a few dozen more in others.

It was not unusual for me to spend part of my morning walking around the building, checking in with my employees, all in the name of being available. I didn't know it at the time, but I was essentially waiting for an urgent issue to find its way to me. When it did, I would settle in and finally get to work.

Of course, it was in those moments, just as I began to focus on a task, that employees would walk into my office to ask for a moment of my time.

Since I had an open-door policy, I would nod my head. They sat in the chair on the other side of my desk and began to share their pressing issue that required my support.

More often than not, these issues *were* important. And I wanted to hear them and support them.

If I'm honest, I welcomed those interruptions despite the fact that I had just begun working on my own important projects.

Here's the problem with an open-door policy: it allows interruptions.

My attention was usually split during those conversations. As much as I wanted to be fully engaged with each person, I could feel the pull of the project I had been working on.

Because of this, I was never giving my full attention to the person in front of me *or* the work I had been doing.

I allowed these interruptions. I justified them in my desire to portray myself as an *always available* leader.

In reality, I was pushing my own projects off and, quite honestly, never giving the person the full attention they deserved.

Of course, as my days would come to a close, the realization of not having been able to focus on *my* work became overwhelming.

Because I let the important work find me so late in my day, I was now pressed for time.

With a 40-minute commute to make it home to my family for dinner, I knew just how late I could push.

Stress and anxiety mounted.

I would hastily cram as much as I could in the remaining morsels of the day. I worked until the last possible minute, grabbed my keys, coat and laptop and made a mad dash for the door.

My desk was piled with unfinished projects that I vowed to sort through the next morning. I usually didn't.

Sound familiar?

Don't get me wrong. I was productive when it mattered.

By most measures, I was successful as a leader. I managed to 'make it all work' for years, yet I knew that there were always better ways to approach my work.

My dirty little secret is, despite being at my desk for 9 or 10 hours, I was rarely productive for all or even most of it.

It's not that I didn't want to be or try to be, it's just that I didn't know how to maintain my focus for that long.

I didn't understand how the volume of small decisions and number of interruptions I allowed (and welcomed at times), distracted me from what mattered.

I didn't put in 9–10-hour days because the job demanded it. I did it because I felt like I had an obligation to account for the ups and downs of my productivity and feel even vaguely good about what I had done that day.

The truth is, at the end of the day, even though I worked hard, I always wondered if I had really accomplished what I was supposed to.

Even more so, the knowledge that there was so much left to do the next day was always buzzing around in my head. And I'd always tell myself, I'd figure it out tomorrow.

Here's the truth.

My failures didn't fix this problem.

Decisions are distractions

Research indicates that for every decision we make, our ability to make subsequent decisions is negatively impacted.[8]

Add to this that the long-held notion that we are capable of multitasking has been stripped of its lofty perch.

It's not that we can't *appear* to be doing many things at once and even convince ourselves, as I have many times, that we can pull this off. It's that we understand this concept differently in light of the research.

We know that our brains are not actively managing multiple tasks simultaneously, but are continually switching between them. This switching has a cost that impacts our performance and carries deleterious neurological consequences.

Decisions are distractions

For every decision we make, our ability to make subsequent decisions is negatively impacted.

Author Michael Lewis profiled US President Barack Obama for *Vanity Fair* during his time in office.[9] The article highlighted

[8] Vohs, K., Baumeister, R., Twenge, J. et al. (2005) 'Decision fatigue exhausts self-regulatory resources — but so does accommodating to unchosen alternatives,' https://web.archive.org/web/20111004053220/https:/www.chicagobooth.edu/research/workshops/marketing/archive/WorkshopPapers/vohs.pdf

[9] Lewis, M. (2012, October) 'Obama's way,' *Vanity Fair*, www.vanityfair.com/news/2012/10/michael-lewis-profile-barack-obama

the deliberate way in which Obama set up his day to more effectively focus his attention and decision-making energy on matters of critical importance in his role.

President Obama shared how he eliminated trivial decisions, such as what to wear and what to have for lunch, so that his decision-making energy would be available for the most important decisions in his day, like whether or not he should invade Libya.

There are other examples of successful people eliminating these types of decisions. Mark Zuckerberg is famous for his outfit of jeans and a gray t-shirt. Steve Jobs too wore essentially the same outfit each day. Fashion statements aside, the purpose of these decisions to wear only one thing is to eliminate the need to expend energy on 'in the moment' decisions.

Remember the story about waking up, grabbing your phone, checking email, and a dozen other tiny decisions we subject ourselves to?

Did it give you palpitations to realize how much time and energy you spend thinking about where things are, what clothes you're going to wear, or what you should have for breakfast?

Each tiny decision drains a bit of our energy. To make the best decisions in each moment, we want to have all our decision-making capacity available to be applied to what matters most in our life and our business.

So, again, the question we must ask is why would we knowingly subject ourselves to this energy drain?

Why would we knowingly keep dozens, if not hundreds, of decisions in front of us, when we can remove them to preserve our energy for bigger, more important endeavors?

What are you working on?

There are days, probably too many, when we don't even know what we're supposed to be working on.

Of course, there's always our to-do list to remind us, but that's the thing. It's always there, looming over us, and it becomes a cobweb in the corner that we know is there, but haven't the energy to get up and clean it. We tolerate it.

There is always this problem of so many 'things' in front of us that each day we encounter moments where we wonder 'where do I begin?'

My prior work patterns were scattered.

I didn't set my course each day. I spent the morning scanning around like some sort of amped-up radar trying to locate the next thing I needed to do.

Only it was rarely what I actually needed to do. My radar was flawed.

I was reacting to every ping, allowing every interruption, and landing on whatever urgent thing jumped into view.

It was not unusual for me to open my inbox and scan for the next thing. I might go through my paper inbox, or walk through our offices.

This is more than assigning a level of priority; it's about making decisions in alignment with what you hope to accomplish.

What is it that we're *really* doing when we jump from Twitter, to Snapchat, to Facebook, open our email, refresh email, go back to Facebook, read an article (part of it anyway), and then go back to Twitter?

We are spending cycles of our energy looking for a safe place to land until we think, 'oh that looks productive, let's start there.'

These software tools and social media platforms are not distractions on their own. They are a means of connection and communication. The problem is how we use them.

We allow them to become distractions in subtle and 'not so subtle' ways.

Each time we receive a notification on our phones, tablets, or computers, we are being interrupted. Even the soft buzz in silent mode or the new message banner in the corner of our screen captures and switches our attention for a split second.

Each switch of our attention is a decision. Each decision is a distraction from whatever you were doing just a moment before.

Handling ourselves in a distracted world requires us to consider the overwhelming number of decisions we face each day in order to eliminate them or at least pare them down.

Eliminating distractions is about minimizing your need to make decisions in the moment.

This requires understanding what requires your attention and when. You get to choose.

This comes from knowing what matters, the actions you need to take, and setting yourself up to do it easily.

Doesn't that sound better than gritting and grinding your way through your day?

This process is continual.

I am refining my own systems for handling distractions all the time.

I was recently talking with my wife during a short break from writing. As I was speaking to her I received two messages on my phone. Of course, they caught my attention and without realizing it, I stopped mid-way through my sentence, leaving my wife waiting for me to finish my thought.

It turns out that I recently added a new Slack channel and hadn't set all my preferred permissions for it yet.

The messages weren't urgent or important. They could have waited and certainly didn't warrant me shifting my attention from a conversation with my wife.

What does that look like?

Where are you expending your *decision-making* energy? Where are you exerting *willpower*, which could be better directed?

What do you want *your* day to look like?

What 'in the moment' decisions are you continually faced with that you could minimize or eliminate altogether?

Are all of the notifications you receive across your devices absolutely necessary? Which have you simply accepted as the default?

PART TWO

The power of simple decisions

CHAPTER 3

●

Put success in your way

I wrote the phrase, 'put success in your way' on a scrap of paper many years ago. I remember where I was. I remember the realization of what it meant to me in that moment. I still have the note.

Put success in your way is my approach for reclaiming my time and attention.

It is based on three core elements, two of which we covered earlier:

1. Willpower is a limited resource.

2. Decisions are distractions.

3. Habits are a powerful force to which we are biologically prone.

Let's review the first two then dig in to Habits.

Willpower is a limited resource

We can all acknowledge our capacity to call upon and exert willpower towards a goal. We also understand that doing so requires a certain amount effort and energy.

With that in mind, it makes sense to consider how to apply our efforts effectively and efficiently. It makes sense to direct it towards areas of your life that matter most to you.

Decisions are distractions

What, where, when, why, and how are not questions I want to consider the moment I open my eyes each morning.

I don't want to start my day walking through my house looking for car keys, thinking about what I want to have for breakfast, or scrolling through notifications on my phone.

Each one of those actions requires a decision, however small.

Each decision takes a moment of our time, however short.

If each moment is taken, it means some other part of our life gets less of our time. Whether that's from sleep, growing our business, or time spent with our children. It all adds up.

Habits are a powerful force to which we are biologically prone

This is the third core element of *put success in your way*.

This is not to say we are completely governed by our habits. However, we are biologically prone to form them.

Habits get a bad rap. Well, bad habits do.

The truth is they play a valuable function in getting us through our day.

It's because of our habits that we don't have to expend mental energy thinking about things such as *how* we brush our teeth or the process of tying our shoes. They are ingrained in our muscle memory and part of our habit bank.

Habits can actually make things easier. They unburden our minds and allow us to move quickly through the day.

Habits present us with an opportunity.

How can we leverage this biological tendency to our advantage?

How can we consistently build habits that serve our needs, making them as simple as brushing our teeth, and using them as a means of focusing our energy and attention?

The foundation of *put success in your way* asks you to consider these three core elements and look for opportunities to leverage your understanding of them to make things easier.

What does it look like for *you* to *put success in your way*?

The guitar in the closet

Shawn Achor's guitar was only seconds away from his usual perch on the couch.

I've never been to his home, but in my mind he could see the closet from where he was sitting.

His decision to place it between the couch and his television *and* remove the remote is a great example of utilizing each of the three elements:

1. He removed the need to exert *willpower* to get off the couch, by placing the guitar within reach.

2. He removed the distraction of making a *decision* between watching TV or playing guitar by storing the remote in the closet. In doing so, he also removed the need to further exert his willpower to avoid watching TV.

3. He leveraged an existing *habit* of sitting on the couch after work.

This is the power of simple decisions.

This is what it looks like to *put success in your way*.

Getting to today

It's not some sort of magic that I'm better at starting my day. I'll confess it's also not always easy. But it is easier.

Knowing what you are working on and what you need to do does require some preparation.

Remember my description of life in my office, the open-door policy, the interruptions, and casting about waiting for the urgent thing to land on my lap.

That continued for years until a day when something shifted dramatically.

It was during our annual family vacation on Cape Cod. As you might imagine, we had full days of beaching planned. Long days in the sun, digging in the sand and body surfing with my kids were all waiting.

As so many of us do, I brought some work with me on vacation. But I was determined to make sure it didn't interfere with my family time. So, rather than carry the burden of work throughout my day, I decided to get up early and knock it out.

I awakened before the rest of my sleeping brood, made some coffee, grabbed my laptop, and started my day at a small table on the deck in the early morning sun.

I knew I didn't have a lot of time, so in order to enjoy my day at the beach, I made the decision to just get three significant things done. After that, I would briefly check email and wrap it up so I could relax and enjoy the day. Here's *what that looked like* in action:

- Item #1 took me about 40 minutes of solid, uninterrupted work.
- Item #2 took me another 40 minutes or so of the same.
- Item #3 required about 20 minutes.

I checked email once, responded to a few things, and closed my laptop. This took about another 20 minutes.

As I was finishing, other members of the family began arriving at the table with bleary eyes and gripping mugs of coffee.

I was able to get all three projects done.

I checked and responded to a few emails. And all in time to enjoy breakfast with my family and help them prepare for our day at the beach.

We were there for the entire day. I never checked in again. After all, I was on vacation.

As I sat on the beach with my wife, I noted that I had done more in those two hours than I often do in an entire day working at my desk.

It worked so well on day one, I repeated this process the next day and the day after.

I chose my projects the night before.

I woke up, knocked out the work, did my daily check-ins, and moved into vacationing with family.

Each day, I was finished, relaxed, and confident because I had accomplished what needed to be done.

Returning from vacation rested and accomplished was an unusual experience, to say the least.

Building upon my success

For years I've had weekly calls with my good friend, Becky McCray. We coach each other, use the time to discuss what we're working on, and talk through ideas.

The value of talking through ideas in this way cannot be overstated.

It is a form of *reflective practice*, which is a deliberate effort of looking back at a process to allow for deeper understanding and the opportunity to truly learn from an experience or observation.[10]

To paraphrase educational philosopher John Dewey, adults don't learn by doing, they learn by reflecting on what they have done.

Becky and I talked through my experience of working on vacation and why it worked so well. We broke my approach into five components in order to replicate the process and refine it into a consistent approach.

I am forever grateful to Becky for teasing these ideas out of my brain.

Here are the five components that enabled me to go to the beach feeling accomplished:

1. *Planned* – I planned ahead of time for what I needed to do the moment I entered the day.

2. *Project limited* – Despite other looming, even important, projects on my plate, I limited my focus and attention to three specific projects.

3. *Time limited* – I had a specific amount of time in which I could work.

---

[10] Dewey, John (1998) [1933] *How We Think: A Restatement of the Relation of Reflective Thinking to the Educative Process*, Boston: Houghton Mifflin. ISBN 978-0395897546. OCLC 38878663

4. *Time specific* – I had determined the specific amount of time I would put into each project or task.

5. *Free from interruption* – I put myself in an environment where people were not going to interrupt me. And perhaps, more importantly, I decided not to allow interruptions or interrupt myself by checking email, social media, texting, taking phone calls, or other distractions.

Let's be clear here. I did not cure cancer in those early hours of the day.

It wasn't perfect or neat. I'm almost certain I cheated a few times to check something when I wasn't supposed to.

That said, after years of struggling to figure out a better approach to how I work, I experienced success in a way that, for me, was profound.

And despite my previous experience night after night of dashing home at the last minute and never truly 'learning from my mistakes,' it was achieving some form of success that provided a platform upon which I could build.

Replicating success

By identifying the five components that made this possible, the next step was to replicate it and build on my success each day.

I started simple. I chose to focus on just the first two hours of each day and approach them in the same way, consistently.

Here's what that looks like.

Planned

At the end of each day, choose the three projects to work on first thing the next morning.

At the top of a piece of paper, write the heading, 'Success ='

Write out the three projects. Be specific.

Project limited

The selection process is critical.

Choose to focus your attention on *only* three, pre-determined projects with the understanding that, during this block of time, you are limited to just these three.

Time limited

Set aside two hours at the beginning of each day.

This means that each project is allowed one, 40-minute segment.

Placing this limitation aids me in the temptation to take a mental break. I know there is an end in sight, at which point I can move on to other things.

Time specific

Having established how much time I will devote to each project, and establishing a two-hour block to work on my three projects, I add it to my schedule in the first two hours of my day and guard it with my life.

Free from interruption

This is the hardest for me.

Perhaps the *hardest* things for me to manage are the interruptions caused by my own thoughts and impulses.

I have rules about not checking email or social media before or during this time.

I find it helpful to begin to let others around you know that you are unavailable during this time. Think of it in the same way you would a two-hour meeting.

I've established some rules to use with my family. For example, when I am working on a project, my wife knows I will probably ignore any phone calls from her. However, if she calls again, that's a signal to me that it's important enough to interrupt *anything* I'm doing. That's our rule.

This is what it looks like to *put success in **my** way*.

It makes effective use of the three core elements.

Willpower is a limited resource. The way in which I am applying willpower is to honor the decisions I've made and the rules I've established. I do this to stay focused on what I had previously determined was important.

By writing everything out and placing it on my desk, it helps me to not be distracted or seek other urgent tasks.

Habits are a powerful force. For me, the hardest part of this whole set-up is the act of writing out my day, the day before, but this is being supported by the success I am having each morning. Success informs the habit.

Decisions are distractions. By writing out the three projects and setting up that time according to those five components, I have eliminated *any* uncertainty or 'in the moment' decision making about what I need to do and when I need to do it.

Random thought interruptions

The one interruption I can never seem to completely eliminate is my brain.

It loves to interrupt and remind me of little things and always feels the need to share a new idea, even in my 'do not interrupt me time.'

It sends me little reminders. The 'oh, I have to call...' moments where we suddenly remember something we were supposed to do. In that moment, we stop everything, switch gears, and quickly make the call, 'before we forget.'

Those interruptions break our focus too, and we lose all momentum.

'Let me do this right now before I forget.'

This is one of the most dangerous things we say to ourselves.

I used to pick up the phone, or move into another project I suddenly remembered, and leave the project I was in half done.

I might say 'I'll just send her a quick email.' Of course, whenever I opened my email I would see 10 other emails vying for my attention.

And I'd think, 'well, I might as well do these while I'm here.'

And then the next thought and interruption would occur.

With all due respect to productivity expert David Allen, I've encountered many people who I believe have misinterpreted, taken out of context, and misapplied his 'If it takes less than two minutes, do it now' rule.[11]

In the context Mr. Allen intended, it makes sense. Taken as a general rule however, you end up in a long series of two-minute tasks and never spend time on the important stuff.

So what can you do with those interruptions?

As a teenager, I was taught a specific meditation technique. My instructor gave me a helpful idea that I use to this day, not just for meditation but also in my work. Especially in the first two hours of my day.

Her advice was simple:

> The goal of this technique is to clear your mind. There are, however, thoughts that enter that seem to interfere with that effort.

---

[11] Allen, D. (2001) *Getting Things Done: The Art of Stress-Free Productivity*, New York: Viking

Our natural reaction is to fight against them, attempting to make them go away by force of thought. This actually makes it harder.

Picture a pool of water. The surface is rough, more susceptible to wind. Our goal in meditation is to go deeper into the pool where it is calmer.

As you are going deeper, it's natural that a thought enters. It's like a bubble coming up and disrupting the calm.

Rather than fighting it, let the thought rise to the surface and disappear. Then, gently come back to your meditation.

Maybe that sounds a bit hokey to you. Maybe it makes total sense. Here's how to apply this idea to stay focused.

Thought capturing

Place a blank piece of paper and a pen next to you during your head-down, two-hour stretch.

As you work, you will inevitably have other, unrelated, random yet important thoughts enter your mind:

'Don't forget X is due in 3 days.'

'You should find a new picture for that post.'

'Remember to call your sister, it's her anniversary.'

Our temptation, especially when it comes to 'call your sister,' is to stop what we are doing, pick up the phone, and call her.

Rather than allowing those thoughts to derail your focus and send you down a rabbit hole, jot them on the paper and go back to work.

Having a quick, easy way to capture those thoughts eliminates the broader impact of the interruption of your focus.

Write it now, address it later. Do it after your two hours are up.

It's just like the thought bubble; let it come to the surface, pop (capture) it, and then move back into your work.

This one technique has helped me stay focused. It reassures me that the idea is safe and won't get lost.

It's rare that you will come across a thought that can't be addressed later or requires your immediate attention.

What are you hoping to accomplish?

Everyone has different ideas for what they hope to accomplish. Some goals can be specific, such as a sales goal or a revenue target. Some are more general, such as being more organized or more present. And others could be as simple as not drinking the water.

Brushing my teeth in Ethiopia

The third core element of *put success in your way* is: *habits are a powerful force.*

I experienced this in a very concrete way when visiting Ethiopia. My wife and I were there to adopt our child.

As visitors to Ethiopia, we were more susceptible to the health concerns related to water. As such, brushing teeth and shaving went from being simple routines we don't have to think much about to things that require a bit of thoughtful consideration and, dare I say, planning.

The simple habits we used to keep us healthy every day could now make us sick if we weren't careful. Since we could no longer drink the water from the tap, we had to adjust. We had to create new systems for ingrained habits like brushing our teeth.

It sounds simple enough, right?

It was my second visit in two months. I thought I had it pretty much down. I had been brushing my teeth using bottled water at the bathroom sink. Easy enough. That's all it takes.

Use the bottled water, not the faucet.

The system was simple. Place a bottle of water right next to the bathroom sink. It worked just fine, until the very last day.

As I was finishing, the ingrained habit of turning on the faucet to rinse my brush and mouth was simply too powerful. I drank the water.

In the early days of the trip, the worry of getting sick and my *will* to stay healthy for the trip was so present that I did fine. I always used the bottled water.

On the last day, however, when the fear of getting sick and ruining our trip subsided, I had become a bit too comfortable. And in an unguarded moment, I slipped. After it was too late,

I realized I had rinsed my brush under the faucet and put it in my mouth.

I realize this may sound a bit silly to break down the act of brushing your teeth. But the real challenge was not about brushing my teeth. I was trying to stay healthy. I was attempting to temporarily (or permanently, if necessary) replace an old habit with a new one.

Willpower can only get us so far. I was very conscious of not wanting to ruin our time in Ethiopia by getting sick. As that desire faded, as our time grew to a close, the immediacy faded away allowing my guard to drop and ingrained habits to resurface.

Habits are a powerful force. I can't think of many habits that are more established than brushing my teeth and there are certain environmental cues that support the habit. A sink, a faucet, a toothbrush, toothpaste, a mirror.

Decisions are distractions. In the face of all of the environmental cues that supported the old habit, simply having a bottle of water on the sink was not enough to change it. All of the other elements were still there.

This meant having to consciously remember (and decide) to use the bottle and remember (decide) not to use the faucet: I should have found some way to cover the faucet to signal not to use it, for example.

Eliminating decisions is as important as making them.

This is very much like starting any new habit.

If we want to start running every day, we may start out strong because of the enthusiasm and new found *willpower* to get in shape and become healthy.

As we know, willpower will only get you so far in an effort to achieve a goal or change a habit. You have to think about how to change your environment to support it.

The small things matter as much as the big things.

When we build our own systems, more often than not we tend towards the complex.

The complex may be impressive, but can be difficult to maintain.

In the end, the purpose is for the system to be set up to serve you, not for you to serve the system.

Habits make things simple

It's important to revisit and recalibrate our habits from time to time. In some cases, it may require a return to consistency. Others require more significant adjustments, but never very complicated.

Habits start out clunky. They seem hard at first. After all, we're retraining ourselves to complete a new set of tasks in place of old routines or where no routine existed before.

The goal is always to set up the steps of each habit in a way that makes them almost impossible not to follow and as natural as tying your shoes.

Six simple habits

There are essentially six habit categories overarching almost every aspect of our lives and these can make a considerable change in a short time:

1. Start
2. Finish
3. Eat
4. Sleep
5. Move
6. Connect

That's it. This list sums up the majority of most days.

How you start and finish your day are two of the most powerful habits you can cultivate.

And while they're in no particular order, I'd actually put 'finish' ahead of 'start.' If I end my day well, if I go through the steps in shutting down, I am inevitably setting myself up for a great start.

Eating and sleeping can be trickier habits to alter.

That said, sleep is what sets me up for my start the next day. I use alarms or reminders to tell me to go to bed on time. Eating habits require perhaps the most significant changes, but at the most basic level it just means following the rules I've set for myself. That's pretty much it.

It doesn't matter whether you count your steps, stand up at your desk for part of the day, or train for marathons. Moving is another habit that changes how you feel and how you operate, almost instantly.

I have limited myself to three options for exercise each day. I have a set amount of time in which I do it. I have an alarm set. I have a training partner. I have an accountability check (I have to text a friend if I miss my workout).

I connect with my friend, John, almost every day. I have others I see in person three to four times a week. I call my Mom and my sister. I make connections at work. I've run with my friend almost daily. You could morph this one to mean 'connect with new ideas' as well and build in a reading habit. That's up to you.

There they are:

1. Start

2. Finish

3. Eat

4. Sleep

5. Move

6. Connect

People sometimes challenge me to add more categories, but I've yet to find something that can't be easily placed in one (or more) of the categories above.

Here are a few I've heard:

- Empty the dishwasher every night. *Finish* – Finish your day.
- Read a book a week. *Connect* – Connecting with new ideas.
- Save money. Eliminate debt. *Start* – Start planning for the future. *Finish* – Finish paying for what you purchased.

I love simplicity. We can always make things harder; it's paring down that challenges us.

The categories are useful in helping us frame our approach.

Clear the decks

Clear the decks is a naval term urging deck hands to 'remove or fasten down anything that might be loose on the ship in order to *prepare for battle.*'

The *doing battle* part of this phrase isn't something I find particularly motivating or inspiring. I personally don't want to go to battle every single day and this book isn't about fighting with yourself. In fact, I'd rather avoid the fight if I can.

But with any navy, being prepared for battle is part of the mission. It is part of protecting the nation's interests against enemies. Waging war or engaging in battle may or may not be part of your mission. You do, however, have a mission, even if it isn't clearly articulated yet.

You also have any number of objects in your life, including thoughts, perceptions, and ideas. Each object either serves you and your mission or gets in the way of what you are trying to accomplish.

There's a lot of physical and mental clutter that accumulates in our lives. In short, there's stuff on your deck. Some of it you need. Some you don't.

It gets in our way and prevents us, not just from being successful, but sometimes from even taking action in the first place.

It needs to be cleared to be prepared to take action when necessary.

Decision minimalism

The more decisions we have to make in a day, the less effective we are in our decision-making abilities.

One of the habits many successful people cultivate is a regular routine with the intention of eliminating decisions, especially the most trivial, from their days.

Everything from what to wear, what to eat, which workout to do, what project you're going to work on that day, even strategic business decisions. They all take a toll on our capacity to decide and take action.

The key is to find opportunities to eliminate the most basic and trivial decisions from our lives. The good news is it's easier than you might think to find some simple areas in which decisions are impacting your focus, effectiveness, and, ultimately, your freedom.

Running is hard

Maybe it's just me, but fitness seems like one of the hardest habits to form.

So many of us fall short of our fitness goals; billion-dollar industries are built to capitalize on this.

They are betting you will fail and it's built into their financial models as well as their marketing.

Like many people, I've tried a number of fitness programs. They were all fine. There was nothing inherently wrong with any of them. The common thread of my 'failure' to follow through had very little to do with the program.

I simply did not understand how to *put success in my way* and build my goals around an understanding of what is necessary to develop the required discipline.

The biggest challenge many of us face lies in all of the obstacles we put in our way in the first place.

These obstacles can be physical or psychological but, more often than not, they are intertwined.

I may *want* to exercise each morning. However, if I have to expend any energy thinking about getting dressed and gathering what I need to get to the gym, you can pretty much bet I'm not getting out of bed.

When I wake up with a goal of working out, if I have to look for my gym clothes, find my work clothes, get my bag together, or find my headphones, you can be fairly certain I'm not going to make it to the gym.

When I was running, if I couldn't find a decent pair of socks or my running shirts were dirty or not where I needed them to be, not only did I not go, I'd roll over and go back to sleep.

Call me weak or unmotivated, but the desire to run was not enough to push through all those silly hurdles.

Don't get me wrong, there are some very compelling reasons for each one of us to commit to some form of consistent exercise.

I applaud you or anyone who gets out the door each day to make that happen. And it is entirely possible for me to have pushed through what are, really, very small obstacles.

That said, they shouldn't be there in the first place.

How I fixed it

Maybe you already know how I solved this problem. You're thinking, 'this is so simple,' 'I get it,' and you're right. You're smart and it makes perfect sense.

To start running consistently, to get up every morning and get outside before I could even think about it, I had to figure out what it looked like to *put success in my way*.

For me, the obstacle to running consistently was not the act of running itself. The challenge lay in all of the steps it took for me to get out the door.

So that is where I focused my attention first. I thought through all of the decision points of the morning. Where in the process of getting up, getting dressed, and getting outside did I hit a snag?

For me, a snag typically happens when I have to make a decision in the moment about anything that could have been decided earlier.

What should I wear?

Aside from making sure I have my clothes ready, I live in the US and in Maine so my clothing choices are weather dependent.

A simple 'Hey Google' or check with Alexa the night before should be enough to make sure I get the proper attire ready for the conditions.

What should I bring?

I like to listen to music when I run. I also like to keep track of my exercise. So I make sure my phone and headphones are charged. My Apple watch is also charged. I use an armband for my phone. I currently don't run far enough to warrant the need for hydration, but if I did, I'd want that ready and available to grab and go.

How long will I run? Or where should I go?

This is a tricky one that many people miss. At this moment, I lean towards an amount of time, rather than a distance. Both approaches are fine. But here again is the issue of a simple decision point that I'd prefer not to waste time or energy making in the moment. It helps to have a pre-determined route and/or an amount of time.

Of course, all of the preparation starts the night before.

Check the weather. Choose the proper clothing. Lay out the shorts, shirt, socks, and shoes. I even untie my running shoes. Make sure the phone, watch, and headphones are charging. Schedule the time. Decide on the route and place everything right next to the bed.

Now, I can get up each morning and simply run. Why? Because I don't have to think about it. It's no longer necessary to go through the tedious process of looking for what I need to accomplish the simple act of running.

Notice the phrase 'act of running.' It's not a goal to run. The goal might be improved fitness, weight loss, or to complete a half-marathon. It is not running. Running is the action, the first action on the way to accomplishing a goal.

When thinking about how to *put success in your way*, the focus is not on the ultimate goal.

You are removing obstacles (all the steps required to get ready) and placing everything you need in front of you. You're making the first action required to accomplish your long-term goal easy and almost automatic.

This is what Shawn Achor was doing when he placed his guitar between the couch and the TV.

Preparation and rules

Our brains do not do well when we try to focus on more than one thing at a time. We are simply not equipped to multitask, at least in the way that we like to think. We cannot work simultaneously on two things.

Psychologist Glenn Wilson demonstrated in his research that multitasking can decrease our effective IQ by up to 10%.[12] Furthermore, increases in the stress hormone cortisol have also been linked to multitasking.

As we succumb to that urge, we suddenly find ourselves *appearing* to do many things at once when in reality we are jumping from activity to activity.

This constant switching on and off is actually bad for our brains and our productivity.

The most effective cures I've found for killing this tendency to 'multitask' are *Preparation* and *Rules*.

Preparation

It's easy to say that you should be prepared. We get it. We acknowledge the concept as generally good.

At its core, preparation is about deciding *before* you have to decide and having what you need available.

Every email we receive, every text, phone call, social media post, like, or reply forces us to make a decision; no matter how small, it is still a decision.

---

[12] Wilson, G. (2005) Info-mania, King's College London, www.drglennwilson.com/Infomania_experiment_for_HP.doc

Preparation, deciding what you will be working on ahead of time and committing the time, means letting go of the constant hum and worry about whether you're working on the right thing.

It means accepting the fact that you can only do *one* thing at a time.

My preparation occurs when I choose the three projects that I am going to work on. I also choose the amount of time that I will work on them.

Typically each project is about 40 minutes in length.

I set the time that this will begin and end.

I also use a timer for each project.

So when the day begins, I don't wonder what I will be working on. I don't try to remember what I have to do. I don't look at my list and decide what I *should* do or what I *feel like* doing in the moment. All of this has been decided ahead of time, which allows me the opportunity to get to work.

I do this with my fitness as well. I have a very specific plan that I am using to accomplish a goal of a successful triathlon.

Each day, my job is to simply execute the plan. It isn't to wonder if what I am doing is the right thing or if I *feel* like doing something else. I simply need to get myself into the pool, out on the road, or, in the case of a New England winter, on a bike trainer.

Rules

We are all presented with opportunities.

What allows me to say *no* or *yes* quickly are the rules I have established in my life and business.

A simple daily example for applying rules are requests to connect on LinkedIn.

We all receive these notifications. Sometimes it's people we know. Many times, it's people from the same industry, colleagues, or competitors.

And you can certainly say yes to every request or you can set up simple rules that make it easier to quickly decide who you'll accept. You can choose to grow your connections intentionally based on your business goals.

Rules from priorities

How do you want to spend your time? How do you serve your customers? What will you do or *not* do in exchange for compensation?

Some of this you may do automatically. The goal is to do it intentionally and consistently to make decisions that will ultimately help you grow.

Establishing my own rules has probably had the biggest impact on my success. Actually, that's not entirely true. *Following* them has made all the difference.

Here are a few of my rules as they relate to *Focus:*

- Do not check email before I have completed my *Prepared* project time. *From the time I arrive at my desk*

until I have completed the three 40-minute projects I decided upon the night before, I do not check email.
- Do not check social media of any kind.
- Do not answer the phone or listen to voicemails.

Those are a few of my rules for staying focused during my early morning. I *do* check email regularly throughout the day, but usually for a limited time and I go in looking for specific communications, not scanning for (or replying to) unwanted distractions.

In addition to those rules, I have a few rules to *allow* for certain interruptions. For example, if my wife calls me and I am working on a project, I 'ignore' the call. She knows I block out interruptions. However, if she calls a second time, I pick up the call. That is our *rule* for communicating something important or urgent. Otherwise I simply call her back when I am finished.

One of my *most important rules* is this:

> My time and attention should be focused on what I have previously determined to be the most important activity at that moment. If I give myself time for a workout (or anything else for that matter), I will not feel guilty about what I am not doing. I am simply honoring my commitment to myself and to the people I serve.

Set simple rules

I can never remember if it was Ron or Paul who first said to me, 'If you do it more than twice, it needs a system.'

Think about all the repetitive tasks you do each day. Do you have a simple system for them? Do you have a set of rules you follow consistently to get them done?

As with everything else, establishing a system and rules starts with what works. Ask yourself:

- What action works for me all the time?
- How can I frame it as a rule?
- What will help me follow my own rule?
- Do you have a sales process you follow every time?
- Do you have a rule for your calendar entries?
- Do you have a rule for the type of client you will work with?
- Do you have a rule or process for screening clients?

Traditionally, rules begin with words like never or don't. For example, I had a rule to 'Never fire someone on a Friday.'

The reason for the rule is, as an owner, I like being available to my other employees the next day to gauge reaction, support them, and set the right tone going forward.

I might reframe my rule from 'Never fire someone on a Friday' to 'Always be available to your staff the day after you fire someone.'

The distinction between something that asks for our attention and something that warrants our attention is critical.

The ability to shift our attention away from the demands of a thousand tiny distractions and purposefully direct our energy and actions is perhaps one of the most important skills we can cultivate.

Put success in your way simultaneously acknowledges our limitations and leverages that understanding to our advantage.

Eliminating decisions, external drains on our willpower, and cultivating habits is the foundation.

Based on this understanding, we can develop simple, repeatable systems that work in the context of our lives.

Rules are there to support how we use these systems and they, too, are our own to make.

What does that look like?

Sometimes the best source for accomplishing a new goal is to look at your prior successes.

Were there examples of how you *put success in your way*?

Did you have a system in place that you could adapt or use again? Did you establish rules for yourself?

Look at your day, at work as well as at home.

View your day in all its various parts and ask, *what do I want this part of my day to look like*?

How can you eliminate interruptions and give more attention to what matters?

CHAPTER 4

●

You are the architect of your system

There are two key parts to the sentence 'You are the architect of your system.' We'll discuss them both separately and focus on how they work together.

The first of these is the word *architect*.

The second is *you* and, more importantly, *your* system.

The word architect has its roots in the Greek word *arkhitekton*, which – literally translated – means, 'chief builder.' It is a great reminder of the ownership we have over our decisions and actions. We are in charge of building the experience we want.

Within the confines of your day, your schedule, your choices, your home, your desk, your desktop, your software tools, and even the way in which you choose to place the apps on your phone, you are the architect; you are the chief builder.

Architects are intentional.

Their designs have purpose.

And while we often think about the physical buildings themselves, the objects they design, an architect guides us and shapes how we utilize or experience the space inside and out.

At their best, designs are meant to evoke and even inspire.

In the book *Nudge: Improving Decisions about Health, Wealth and Happiness* by Richard H. Thaler and Cass R. Sunstein, the authors discuss 'choice architecture':

> A choice architect has the responsibility for organizing the context in which people make decisions.[13]

Examples of choice architecture are all around us in our everyday lives.

Every website we encounter and the various apps we use are designed with the intention to help us make a decision. Of course, as helpful as that may sound, the decision is not always for our benefit.

Menus are another example of choice architecture. Have you ever noticed a restaurant menu with specific items circled?

That simple visual cue is designed to increase sales of those items. And it works. As it turns out, those items typically have the highest profit margin for the restaurant. What seems like a helpful cue may not be helping us at all.

---

[13] Thaler, R.H. and Sunstein, C.R. (2009) *Nudge: Improving Decisions about Health, Wealth and Happiness*, New York: Penguin Books

Another great example of choice architecture we encounter in our everyday life is through product placement.

We may not think much about it, but grocers and retailers are well versed in the power of choice architecture and they are intentional in how they 'feature' items to entice buying.

Even when faced with dozens of options for pasta sauce, grocers (and the makers of the sauce) know that placement on certain shelves (generally the ones at eye level) is key in getting people to buy a product.

Good or smart design funnels our decision making. The design itself can often limit our choices. Even when there are dozens of options, the right architecture can narrow them down.

Done well, with a goal of service to the customer, this can be incredibly helpful. But every day in small but significant ways, we allow ourselves to be guided by the architecture others have created.

We accept the default settings without considering that we may get to choose something different.

There are strong forces at work to influence our decisions and, if we're not careful, we can find ourselves drooling in the pasta aisle trying to choose a simple sauce.

If you don't do it, others will do it for you

We hate to admit that we are susceptible to this, but all evidence points to the fact that choice architecture is a successful strategy for influencing our buying behavior.

Information like this is powerful for a number of reasons:

1. Knowing that nearly every experience we have in our lives is impacted by someone attempting to design the choices we make allows us to see the experience for what it is. We can appreciate it, or at the very least enter it willingly. That in itself is a choice.

2. We can leverage this knowledge for our own efforts. Using our understanding, we can design environments and experiences that help funnel our choices and direct our attention towards our chosen goals.

A simple and proven example of building a system to avoid or at least reduce our susceptibility to the influence of product placement is to enter the grocery store with a list and a budget.

Research shows that when you know what you are going in for and you have a set budget, you can more effectively eliminate those influences *and* reduce your grocery bill.

Systems don't need to be complex to support us. To focus our attention on the choices *we* want to make, sometimes it's as simple as making a list.

Put success in your way is an example of choice architecture.

Setting up my day the night before with the three most important things that I need to accomplish, written out, helps support my experience as I enter my day.

The blank piece of paper next to me as I write is a place to put the ideas that enter my mind, not unlike a set of well-placed shelves, 'a place for everything…' so to speak.

We have to ask (and answer) the question, *'What would that look like?'* with regards to the various points in our day.

'Of your system'

Look at those words again.

You are the architect of your system.

At first glance, we get it. We nod our head and, on premise, agree with the idea of taking back control. But if we simply think about the first part of the sentence, of being the architect, we miss the point.

The second and equally important part of the sentence is 'of your system.'

Maybe something isn't working for you.

The reasons may be long or short, but many have involved some form of prioritization, procrastination, organization, or distraction.

Interestingly, this list of reasons why it's not working always seems to come more easily to us and we rarely, if ever, can list the reasons why something worked for us.

It's important to reflect on your success as Franco did to identify a successful sales process.

The point of the exercise he did, however, wasn't some warm and fuzzy 'don't forget that you're special and great' rah-rah.

It is rooted in the idea that we so often ignore the fact that we have built our own systems of success. And they are systems we can replicate, adjust, and build upon to address future challenges.

Put success in your way is not meant to be a specific formula, but a framework to approach your specific challenges.

The point of examining your success is to realize that you probably have existing frames for achieving your goals.

There are methods that have worked for you in the past, elements of which can be used to achieve success again.

When I ask people how they did something successfully, it is often met with an uncomfortable silence, a shrugging off of the question, a few may sheepishly mutter comments like, 'I don't know, I just did it' or 'I work hard.'

However, if I ask that same person why they failed, the list of reasons suddenly rolls off the tongue.

This would be fine except we don't always close the gap on the reasons we failed, and when we do, we usually patch it over rather than reflecting and learning from it.

By starting with prior achievements, we start with demonstrated evidence of success. The goal is to understand and adapt it to new situations.

Framing your plans

For years I fell victim to believing that once I failed on an effort (exercise plan, course, resolution, etc.) then it was over.

In crept the negative voice in my head saying, 'there you go again, you know you never finish *anything*.' I let that voice win, often.

Sometimes we need to give ourselves permission to start over. When that fails, sometimes we need to hear it from someone else. With that in mind, I give you permission to start over, right where you left off.

I'm guessing hundreds of ideas pop into your head each day. What do you do with them? They can appear almost out of nowhere but how many ever get the attention they deserve?

With all the distractions in our lives, we don't always take the time to consider them. Even our best ideas get lost. We either act impulsively or the idea drifts off and nothing comes of it.

How do we get from idea to deciding to take action?

What does that look like?

You know that I jot down ideas and random thoughts throughout my day on a blank page I keep next to me.

When I make time to review the sheet, I need to decide what's next for everything I write down. Sometimes it's simple. Other times it requires a bit more thought and I'll make time in my schedule for that.

My favorite question to ask when fleshing out an idea is always 'What does that look like?'

It is one of the most useful starting points because it can quickly move you from just a concept to developing a plan. That question helps frame up the steps you need in place or actions to take to make the idea real.

Take any goal you have or any success you are hoping to achieve. Ask yourself 'What does that (process) look like?'

In almost every business conversation I have, this question has come up. It helps us to make better business decisions by thinking through the steps.

We take an idea and talk through what it looks like.

It can turn an idea into an action plan or, in some cases, it can help us reach a decision not to pursue the idea at all.

And then what happens?

In any good story we want to know what will happen next. There is a sequence of events unfolding that will (hopefully) bring us to a conclusion.

When you ask 'What would that look like?' you begin to think through the sequence of action and how it will unfold.

There are two other questions that will help you lay out a plan:

- What happens next?
- Does anything *need* to happen before this?

Oftentimes the first step you identify in a process isn't always the first step. Asking these questions helps to frame the process from many sides.

You quickly end up with a simple series of steps for putting your idea into action.

In virtually every course or webinar I have created or offered, I've used some version of this process.

It's not just helpful for creating and launching.

It's also the framework I use to walk *you* through the process. For example, in one course we actually show you what the process of *making an online course* 'looks like.'

Try it yourself. Take one of those great ideas you have, a problem you need to solve, or even a goal you want to achieve. Ask yourself '*what would that look like?*' and see how quickly you come up with a series of actions that can make it a reality.

So, what did your path(s) to success look like? In starting your business, getting a job, losing weight, what were the steps?

Do not include things like:

- I have a strong work ethic.
- I just did it.
- I decided to.

Look a little more closely and think about your success through the lens of:

- *Willpower* – What motivated you to start in the first place? What made exerting willpower in that situation appealing?
- *Habits* – What one, two, or five things did you do consistently to make that success a reality?
- *Decisions* – What decisions did you eliminate to ensure that this was what you were focusing on?

Thinking this through a bit more, let's look back at the five components:

1. *Planned* – I knew what I needed to do when I entered the day.

2. *Project limited* – Despite other projects I might have had going on, I limited my 'day' to this one.

3. *Time limited* – I had a specific amount of time in which I could work.

4. *Time specific* – I had a general idea of how much time each item would take.

5. *Free from interruption* – People were not going to interrupt me and I also knew that I couldn't interrupt myself with Twitter, Facebook, etc.

If you boil down those five components a bit further, you will notice three main categories:

- *Time* – Two of the five are related to time limits and time specifics.

- *Choices* – Two of the five are related to limiting choices and focusing only on those options.
- *Distractions* – The last is rooted in eliminating distractions and interruptions.

Here's the real work...

Using the frames above, describe what *is* working in your life. What *are* you able to accomplish each and every day, and how does it line up with the five components?

Don't worry if it doesn't seem to fit exactly, I just want you to think about it in those terms.

Pick two aspects of your day that you would like to improve upon. Here are some ideas:

- Email – Overwhelming and too much of my time.
- Social media – It's part of my job, but I find myself down rabbit holes and justify two hours by saying, 'well, I read some interesting articles.'
- One of the projects for launching your new business – I have to launch the website, I have to set up the ecommerce, I have to write contracts, I need to build my new social media strategy.

Build a frame for how you will address them based on the five components:

1. Planned

2. Project limited

3. Time limited

4. Time specific

5. Free from interruption

It might look something like this.

Rob's email frame

1. Email is the first thing I will do *after* my two-hour 'Success =' time.

2. Respond to email in batches based on who they are from (this may require setting up filters).

3. 15 minutes each time maximum.

4. No more than four times per day. At approximately 9:30/11:30/1:30/3:30.

5. Rules:

    •   Email is not real-time chat. Respond and move on.

    •   If it takes longer than two minutes, it's a task not an email. Move it to your list.

    •   No checking social media and no phone calls (this one may not work for everyone).

Keep that blank piece of paper next to you for this too. This is still focused time and you need a place to put those ideas when they bubble up.

A few notes about this:

Before getting too specific about how I built and use my email system, the point of this example is to demonstrate how I might take the goal of approaching my email differently.

I accept my role as the architect of my system, and apply the framework that helped me be successful at the beginning of my day to other aspects as well.

Your years of experience have given you a foundation of expertise, not only in a specific field of interest, but also expertise rooted in what works for you and what doesn't.

If you're a parent, for example, you've probably learned a few things that may benefit a new parent.

If you've worked in an organization for several years, you can help a newer employee navigate the environment.

As a business owner, you've probably weathered some ups and downs that you could help others avoid.

Perhaps you have some 'if I only knew then, what I know now' perspective on your life or your business.

Others could benefit from this knowledge. Of course, you are more than willing to share your advice when someone asks, or you're required to assist.

What could you ask yourself? What advice are you currently seeking that you may already have an answer for?

What's your problem?

Is it a particularly challenging time right now? Do you have a handle on it, or would you appreciate some help?

Here's the thing, you already know how to start solving it.

If someone came to you with the exact same problem, you would be able to generate ideas for how to approach it.

Your experience and your previous successes or failures have taught you something. You understand what works and what doesn't.

You have methods or patterns that you use. Some help you succeed, and you rely on them. Others get you stuck so you know to avoid them.

So, if someone came to you and described your situation, how would you advise them?

How would you help them approach the problem differently based on what has worked for you?

Follow your own advice

It is easy to support others, but much harder to take advantage of our own wisdom.

Maybe it's because we don't like the answer or maybe we don't trust it.

Part of our responsibility when seeking advice is attempting to solve the problem in the first place.

I'm often surprised how when I stop and look carefully at a challenge, the solution is right in front of me. I don't always see it because I'm so caught up in things 'not working' but it's usually there.

Here's an exercise:

Start a new email. Address it to yourself. In the subject line, write: 'Need Your Advice.'

Write out the problem or challenge you're facing as though you were asking a trusted friend and advisor.

Send it to yourself.

Open the email and respond as though it were your best friend or a client reaching out to you for support.

If need be, make it a conversation. Ask for clarification.

For example:

> Hey Rob – Sorry to bother you, but I really need your advice. My new business just isn't making enough money right now for me to quit my full-time job. I feel like I've tried everything to make it work and I'm getting tired of working two gigs and never getting ahead. It's causing some stress at home and I just don't know what to do next.

If a friend of colleague reached out to you with *that* email, I bet you'd have something to say.

Moreover, I bet you'd want to ask them a few questions to get a better sense of the problem.

Maybe you'd ask about the sources of revenue, and their sales process. You might want to know what they've tried up to this point.

Maybe you'd ask about the stress at home or whether they've established a revenue benchmark for when they could leave their full-time job.

You might sense them feeling overwhelmed and offer to help them to break things down into more manageable chunks.

The point is this: you would also make time to support them. You might even schedule a phone call or talk over coffee to help them think through the problem. You would offer ideas for what to do next.

Just like any conversation, in the end, the advice you give them may not provide the full solution. That's okay.

What it does though is help them reframe their challenge. Asking questions helps them see their options more clearly and puts them in a better position to make a decision about what to do next.

We need to give ourselves the same time and thoughtful attention. We should honor our own needs and challenges in the same way. And if the advice we'd give to others is sound enough to share, why wouldn't we trust it ourselves?

I know you have it in you, so don't be afraid to ask. More importantly, make sure you *do* what you tell yourself to do.

Reflective practice

It helps to ask the right questions. What are the *right* questions?

I'm glad you asked. I'll give you the short answer.

The *right* questions are the ones which, when asked, result in an answer that moves you forward in your efforts to be better.

I get that that's a bit of a non-answer answer. It is, however, absolutely true.

To get to the place where we ask ourselves the *right* questions, it is important to understand how we learn as adults.

When John Dewey encouraged learning through reflection, he was speaking about adult learning, specifically.

Children learn by acting on their world. Adults require some amount of reflection on what they have done in order to learn from the experience.

Reflective practice is a concept introduced in 1983 by Donald Shon. That's 50 years after John Dewey made note of it's importance.

There are a number of definitions for reflective practice each based on the context in which it is applied.

For our purposes, let's use the following definition:

> In reflective practice, practitioners engage in a continuous cycle of self-observation and self-evaluation in

order to understand their own actions and the reactions they prompt in themselves and in learners (Brookfield, 1995; Thiel, 1999). The goal is not necessarily to address a specific problem or question defined at the outset, as in practitioner research, but to observe and refine practice in general on an ongoing basis.[14]

I was particularly drawn to a couple of things in this definition.

Continuous cycle

Taking time to review your work and reflect on it has to be a regular occurrence, a consistent practice.

Not specific

The 'goal' is 'not to address a specific problem' but to 'observe and refine practice.' For our purposes, this works. You want to get better.

Productivity icons such as Dale Carnegie, Stephen Covey, and David Allen all have as part of their teachings and methods some reference to taking time to do a 'weekly review' (Allen) or encourage you to 'sharpen the saw' (Covey).

In this context, I see it as an important tool in the process of creating space between stimulus and response, and directing your attention to the work that matters.

---

[14] Florez, MaryAnn Cunningham (2001) 'Reflective teaching practice in adult ESL settings' in *ERIC Digest*, ERIC Development Team, https://files.eric.ed.gov/fulltext/ED451733.pdf

The *challenge* though is that it's one thing to accept this as a premise, it's quite another to be able to put this into practice.

Remember my favorite question?

What does that look like?

What does reflective practice look like for you?

- 'That went well.'
- 'Nice job.'
- 'That was helpful. Thank you.'

These are all responses I received after I presented a workshop to a small group of people.

It *did* go well. I know the information was helpful. People left with good information and ideas. Then I left the room and drove home.

In the quiet of the car ride home, their words and their reactions became hollow.

I started to realize that I missed the mark somehow.

Sidenote (and a warning): many, if not *all* of us, have some amount of post-performance insecurity. To alleviate that, we look for the feedback and reassurance of others who 'experienced' us to give us perspective.

However, most folks are unwilling to be critical, let alone constructively critical, with us. Some of this is part of our cultural norms. More importantly though, it's not their role.

My advice is not to dive too deeply in self-criticism and to simply be grateful for the positive reactions without giving too much weight to them.

Get some perspective in the form of distance and then evaluate.

So, how do we use reflective practice to get better?

How do we accept the positive reactions and acknowledge our shortcomings without giving too much energy to either one?

Actions and reactions

Our internal dialogue tends to be dichotomous, as is displayed in our evaluation of our performance.

We generally focus on two questions: What did I do wrong? What did I do well?

What's more, we tend to focus on what went *wrong* disproportionately. We believe, and we are not entirely wrong, that if we focus on shoring up our weaknesses, fixing what is broken, we will improve.

As I said, it's not wrong to think this way. Our weaknesses deserve attention. Our mistakes require adjustment.

However, the answers we are seeking to get better rarely lie in what went wrong or what isn't working. They lie in asking '*what went well?*' and, more importantly, '*why?*' and '*how do I replicate it?*'

We need to move from exhaustively examining our failures to recognizing them and moving through them. We need to focus our efforts on what went well, knowing what success looks like, and understanding how we can replicate it.

In their 2010 book, *Switch*, brothers Chip and Dan Heath tackle the subject of change. More specifically, as their subtitle suggests, *How to Change Things When Change Is Hard.*

One of the topics they tackle is our obsession with training to the problem, rather than the solution.

Our brains are quite adept at finding problems, particularly when we are working towards a goal. We can find all manner of things that we do wrong. We can get locked down in a cycle of negative speak and disparage ourselves in ways we would never disparage others.

The Heath brothers not only cite examples of individual change, but also of large cultural change made not by focusing on what a large portion of a population *wasn't* doing, but on what the minority of people who are successful *are* doing, the 'bright spots.'[15]

Highlighting what works is more effective. Using the people, their actions, and methods as examples of the behavior we would like to see made the change easier.

If you're looking for a job, is it helpful to read a list of what *not* to do, or is it helpful to follow in the footsteps of people who have been hired and replicate their methods?

---

[15] Heath, C. and Heath, D. (2010) *Switch: How to Change When Change Is Hard*, New York: Broadway Books

If you're improving your fitness, is it helpful to have a coach point out your bad running form or the flaws in your swim stroke, or do coaches instead help you to understand the proper form and attempt to replicate that?

Going back to my presentation, I left with a general sense that everything went okay, but I began to feel as though I didn't land the messages in the way I had hoped.

Upon reflection, there were times when I noticed information falling flat or when people were less engaged. While most walked away with something valuable, I could tell that I failed to enable a change in their behavior going forward, which was the goal of the presentation.

As you can see, I am focusing on what didn't go well.

Again, this isn't wrong, but it doesn't necessarily solve my problem. It doesn't help me improve or change my approach next time.

What happens if we shift where we look to gain insight for improvement?

- What part of the presentation went well?
- What were you doing or saying?
- What did you notice about the audience?
- Why do you think it went well?

The goal of these questions is to help me recognize what works, in order to replicate it in the future.

Let's use that framework to reflect on that presentation.

What parts went well?

I felt as though I connected with the audience when I was most comfortable with the material.

What were you doing or saying?

It was during the three sections I had presented the week before to another group. I also was being very specific, providing an almost step-by-step instructional process.

What did you notice about the audience?

During those moments, people were more engaged, asking more questions, and taking notes as I (and others) spoke.

People were also more attentive and relaxed when I used personal examples to introduce new ideas, finding ways in which to contextualize the information.

If I were presenting the same information again, I would:

- Review and rehearse the newer material.
- Use the framework that seemed to resonate most:
  - Concepts
  - Personal stories illustrating the concepts
  - Step-by-step instructions or actions

In reflective practice, I choose to focus on questions that ask what went well.

To understand better how my actions influenced that, I ask observational questions such as: what I said or did, what I noticed as a reaction on the part of the audience, and what I thought went well and why.

Through that, you can come up with a simple framework, and a plan for the next time:

- In order to improve, we must ask the questions that help us become better.
- In order to ask the right questions, we need to allow for the time and space.
- Implementing a habit of reflective practice, where we can examine our performance, is critical to growth.
- We have a tendency for dichotomous (what went wrong/what went right) thinking.
- We give disproportionate weight to our failures.
- Focusing our attention (and questions) on our successes provides us with a framework for replicating success in other areas.

Be your own assistant

The value of a good assistant is difficult to overstate. Having someone preparing your day so that you have the information you need right when you need it would make life and work so much easier.

How would your work day be different with that level of support?

Schedule and direction

'You have a meeting with…' and 'This way, sir/madam' have to be among two of the most common phrases uttered by assistants supporting executives of large organizations or even countries.

Their schedules are often set well in advance. The destination and locations are determined. There's no need for them to think much about that level of detail leaving more time for important decisions.

I think many folks reject the notion of a day scheduled out like that. But isn't that what so many of us are struggling with?

We say we're busy, but I'm betting we're overstating that. Our primary problem is more with the decisions and distractions of what to do next, and where to focus our attention.

Personally, I imagine it to be quite freeing (if not somewhat demanding) to have these things nailed down. The idea of an assistant preparing the information I need for my day and laying it out in my calendar is quite appealing.

Travel information, notes and contact information for a meeting, a sales call sheet prepared with who to call, their numbers, and a note about each person. All of it laid out in order to move from one important task to another throughout the day. And always leaving with the feeling that you worked on exactly what you intended? It sounds fantastic.

You and I may not have that level of support but that doesn't mean we can't take some cues from how to operate this way.

Decide before you have to

I work to eliminate as many decisions from my life as possible. All those tiny little details about what to wear and what to eat get in the way of the bigger stuff. Deciding ahead of time, making sure everything I need is ready and available creates a much easier morning.

Fill your schedule

While I don't pack my schedule full of meetings or calls, I do schedule all of my time during my day in order to be more deliberate with how I use my time.

But I schedule my free time so that I can make better use of it, even if that means scheduling a nap or a walk in the middle of the day. It's far better than filling unscheduled time with random internet browsing and scanning social media.

Use the tools you have

Your email has filters that can deliver messages into specific folders. By taking a bit of time to set them up, you can get exactly what you want in each folder.

Your calendar has sections for location, contact information, and notes about your meeting. Filling in this information means no searching around your inbox for emails or phone numbers two minutes before your next call.

This isn't to say it won't take a little work upfront, but imagine what you'd want an assistant to have ready for you.

What information would be most helpful to you the moment you need it? What would your day be like if everything was laid out?

I'm just guessing, but I'm betting it would help you focus your attention on what matters most.

We need time and space to renew and recharge.

I first encountered reflective practice in a formal way through the Brazelton Touchpoints Center.[16]

In that context, it is used as a method of continually integrating and deepening the application of the Touchpoints approach.

In your work, it can be used as a method for improving your approach to making better decisions.

Method and practice

This is part method and part practice. The method is the set of frames that you build to support your work. It is how you *put success in your way.*

The practice happens two ways:

---

[16] Brazelton Touchpoints Center, https://www.brazeltontouchpoints.org/

- The *first* is every day, in your work and your life; it is the act of using the framework you've built to leverage and support your decisions in all aspects of your life.
- The *second* is in your preparation. It is the act and habit of planning out your next day.

Unlike an athlete, musician, actor, or another performer, we are not able to always separate our work from our practice. We are not all gearing up for the next game, match, or opening night. We do not seem to have the luxury of saying that we are practicing today in preparation for something else. Unless we are. Unless we do.

The work is the practice

Athletes go to work every day. They show up at a practice facility, pool, or gym every day to do the work that enables them to perform when they are called to. There is a framework that they or their coaches have built for them. They implement routines they know will help them to be successful.

As they go through each routine, working within the frame, they are also making mental notes about what's working, what isn't, and what needs to be done differently.

Their work is their practice. The routines they do each day are part of the 'job' of being an athlete. Throughout their work day, notes are taken, ideas come, and they are reflecting in the moment about what they need to do differently next time.

At some point, the notes have to be reviewed, by them, by their coaches, and adjustments need to be made. New plans are drawn up for the next day.

Does this sound familiar?

Our focus is on success, on what we are doing well. While we don't bury our heads in the sand about our failures, we focus on building on what works because it gives us a certain momentum. Just asking yourself the question, 'What worked for me before?' is a form of reflective practice.

With that in mind, I want to point out how you are already using reflective practice.

Reflecting as preparation

The simplest form of reflective practice can be found in the habit of preparing at the end of the day for your next day. Taking time to review what you accomplished during the day and writing out the next day's plan. It is an act of *putting success in your way*.

There are two tools I use for this each day: the daily sheet and the on-deck list.

My end of the day process is simple:

- I put a new daily sheet in front of me.
- It has space on the top I have labelled 'Success ='

- These are for the three projects I plan to select for my two-hour block of time.
- 'Success =' is the priority.

Below these is my on-deck list.

This is the list of ongoing tasks we all have. I keep them all in one place, on the sheet. I don't use software to manage this. I use my sheet.

Each day, I write my daily sheet out by hand.

If something wasn't done that day, I move it to the next day, writing it out again by hand.

The purpose for this is simple but effective.

If you write the same thing more than a few times, you're going to either prioritize it or discard it as unnecessary.

In this way, each time I write out my daily sheet, I am engaging in a form of reflective practice. It forces me to reconsider the level of importance.

Remember the blank page

I've talked about this as a critical tool for enabling me to maintain focus throughout the day.

The way in which I use it during the day is as a method of thought capture. As I am working on a project, or writing, I jot down my random thoughts that would normally distract me and take me in a different direction.

Sometimes this is something I simply need to remember. However, it is also one of my tools for reflective practice.

This is where ideas for improvements on my implementation of my systems or methods come from.

This is where I capture quick thoughts for new ideas for other projects I am working on.

I can't turn my brain off, but I can manage my impulse to jump from one thing to another.

That is how it can be used in the moment, but the reflective practice comes when I review it.

At best, this is done at a couple of points in the day.

The *first* is at the end of my two hours of focused time. After a short break, of course. If there are reminders that popped up in the two hours that should be added to my daily sheet, I will add them.

The *second* is at the end of the day in preparation for the next day. This is where those notes are added to the sheet or noted for the future.

These two tools provide a simple framework for reflective practice and reviewing, reflecting, and preparing for the next day.

You're doing it already

The concept of reflective practice is important and critical.

Having a coach can aid and support this process. I use my own coach this way.

But it is possible to build this into a personal, consistent, weekly review practice.

For now, keep it simple.

Give yourself time to review and prepare at the end of the day. I promise that once you do small bits of this, you'll want to find ways to do it more and make a bigger effort of it.

Here are a few phrases to keep in mind when coaching yourself through this process:

- What would that look like?
- What did that look like?
- Why did it work?
- Why didn't it work?
- What got in my way?
- What is the first thing I need to do to make this work?

Ask the questions. Listen for the answers.

Measure, dump, and refine

Ron Hood is a force of nature. I've worked with Ron in some capacity longer than anyone else in my entire career.

I first hired him while at a non-profit organization. He was in charge of recruiting, screening, interviewing, and matching high-school mentors with elementary school children as part of the Big Brothers Big Sisters of America program.[17]

Ron is a natural systems guy. In many ways he makes it look easy. I know it's not. He works hard. But to make it less difficult, he takes time to set up systems that make his life easier.

I've learned a lot from him, especially about how to measure, dump, and refine.

Ron's recruiting story

Each September, as the school year begins, there are a number of events that appear perfect for recruiting mentors. Freshmen are arriving, orientation meetings, add/drop periods, and parent nights are all happening. Ron would attend many of these and set up an information and recruitment table for his school mentoring program.

Ron values his time greatly and *hates* to waste it on anything that isn't getting results.

That said, when he began with a new school, he would cast a wide net, measure the results, dump what wasn't working, and refine the method.

Measure

You can picture the table. Pictures of high schoolers mentoring children, a bowl of free candy, brochures explaining the program, inspiring stories, statistics, and, of course, applications.

---

[17] Big Brothers Big Sisters of America, www.bbbs.org/

These are high schoolers and you can also imagine that many will walk by, ask a few questions, take some materials, and maybe even an application.

I listened to other coordinators assess their night based on an empty bowl of candy. They'd recall with excitement how they talked to 'so many' high schoolers, their voice was hoarse. Seeing a few small stacks of brochures and applications left on the table, they'd count the night as a success. Not Ron.

Ron is different. Ron loves the program. He loves engaging with potential mentors. But Ron also values his time.

Ron would track four simple metrics for each event to determine how successful it was:

1. How many kids took applications.

2. How many of the applications taken were actually returned by the deadline.

3. How many of the applications returned made it to interview.

4. How many interviews made it to match.

He tracked this information in each school district, for each event.

He recorded it in a simple spreadsheet (that eventually grew to be not so simple, but that's a story for another time). Then, armed with all of this data, he set it aside, for now.

Ron continued with the school year, training mentors and supervising matches, and focusing on providing support to each of the matches to help grow strong relationships.

Dump

At the end of the school year, after closing out his programs and matches for the summer, Ron began his analysis.

Equipped with the data of the four results for each event, he determined quite easily which were the most successful for him. He then mercilessly dumped the events that simply weren't worth the effort.

Ron then refocused his energy on events that brought the greatest results for the beginning of the next school year.

It was simple. It made so much sense.

However, he was the first person (and the only person) on the team to approach it this way.

To others it seemed like so much work. Numbering each application, writing down the names of each person who took one. Tracking whether or not that person returned it, let alone whether or not they passed screening, interviewing, and matching... phew!

It was a lot of work, but because he set up a system, because he measured, dumped, and refined, he spent less time on events that weren't producing and had more time to do other things.

While others were content to spend hour after hour, year after year at events never really knowing the results from last year and whether or not it was even worth their time, Ron was:

- Supervising more matches than everyone else.
- Raising more private donations than everyone else.
- More than doubling financial support from funding organizations.
- Being invited to partner on federal grants.
- Being given free bus transportation for his mentors.

Refine

Ron was also an amazing note taker. Over the course of each school year, he would make notes like:

- Need to add X to the application.
- Want to track time to match for each school.
- Add new question to interview form.

What always amazed me about Ron's process was that he didn't make significant changes immediately.

He gathered data and made notes. He wanted to watch his system play out and give it a bit of time.

He trusted the system he set up and let it work. Sure he made some simple tweaks here and there, but no real wholesale changes to the core elements of his system. He just made notes along the way.

And much like the data he gathered on the events, before entering the next school year, Ron would make the changes necessary to perform his job more efficiently and effectively.

Too often we make changes mid-stream without fully understanding the effect or impact.

We make adjustments to systems because they don't *seem* to fit, but honestly what are we measuring? How are we using that information for evaluation? How do we know what to cut and what to add?

Back to measurements

Let's be honest, not every bit of what you're working on will have the same concrete measures as Ron's 'applications-to-match' statistics.

That said, we can be a bit squishy with how we evaluate what works and repeat behaviors or hold on to methods because they've become habits.

What measure, dump, and refine looks like

Here's a question for you. Are you still not able to fully *put success in your way* for the two hours each morning?

Measure

Contrary to what you might think, Ron's story is not about eliminating failures. It's about finding success and using it to grow his program.

Eliminating what isn't working happens as a result of spending more time on what works. He simply doesn't have time to spend on things that aren't working for him. That's not some platitude. I have witnessed this process at work with Ron.

If you've given the two hours of focused work a try, let's set aside for a moment what didn't work for you, and focus on what *did*:

- Were there days that this really worked? Why was that?
- If you were to average out the time that you spent on focused, uninterrupted work, what would it be? Forty-five minutes, 60 minutes, 90 minutes?
- How about selecting your three things for two hours the day before? Did that work better for you on some days than others?
- Did you have more success when you did that?

A summary of your 'what worked' measurements might be as follows:

- Over a two-week stretch I was able to set aside focused time on 6 of the 10 work days.
- I averaged around 60 minutes at a stretch and managed to get through one–two projects each time.
- On the days I did set out my projects the night before, I was able to start better and work longer. Sometimes I did that for 10–15 minutes first thing in the morning, before I started. But 10 minutes the night before worked best.

I will share with you that in the earliest days of applying these principles, this was very typical of my weeks. So, look at it. I mean, really look at what you accomplished.

For the majority of your working week, you woke up to a pre-planned, 60-minute stretch of time in which you were able to accomplish or make significant progress on several 'projects'!

The reason I start with the success you *do* have is to remind ourselves that it did work, in some form.

We can't avoid looking at what didn't work, but this shifts your perspective to one of maximizing success.

Let's take a look at what prevented you from doing it every day. It might look like this:

- I didn't always lay it out the night before.
- I woke up and checked the email on my phone and got pulled into someone else's urgency.
- I opened my browser out of habit, went to Facebook and two hours later I had 'liked' 37 posts and watched two fail videos (hey, it happens).

What prevented you from working for a full two hours?

- I actually finished my three projects in 60 minutes. Go me!
- Interruptions at work. Or I checked the email on my phone and got pulled into someone else's urgency.
- I opened my browser out of habit, went to Facebook and two hours later I had 'liked' 37 posts and watched two fail videos (hey, it happens… again!).

Why weren't you able to lay things out the night before?

- I got caught in a meeting and then had to make it home.
- I don't have one place or a form where this happens.
- I forgot.

In looking at *all* the above, most of this is rooted in one or more of the three core elements.

Willpower

Willpower gets all the blame when we try something new and can't seem to make it work, but it may not be the culprit. After all, you want to be better. I know this much. It's why you've read this far.

Habit

This is where you need frames the most. In order to create the habit, the framework of your system has to be there to support you.

If you had to look for your toothbrush and toothpaste every day, you'd get a bit frustrated and your willpower would cave. But no, you have them stored in a cup or hanging next to your bathroom sink. Those three components (willpower, habit, decisions), plus the fact that the bathroom is the first place you visit in the morning, are your framework for the habit. It is how you *put success in your way* for brushing your teeth.

Your problem with the first two hours might be the frame. It's not stable enough… yet.

Decision

Are there still too many choices?

Staying with the toothbrushing analogy, there aren't many other choices at the sink in the morning.

So (and this is where using your measurements comes in), what did you notice yourself doing before or during those two hours that prevents you from using them as you hoped?

Are you still being drawn to your phone to check email? Do you still have a browser and several tabs open? Does your mind start wandering, land on something you forgot, and pull you into another direction?

Dump

So, what will you dump?

In this case, you're still trying to dump old habits, interruptions. You are still trying to dump black holes of time that pull you off your game. You are trying to eliminate decisions... distractions:

- Dump the phone next to my bed so I resist the temptation to check email first thing.
- Dump the browser. If I must use it as part of my two hours, only open one tab at a time.
- Dump the phone. It's why I have voicemail after all.

Refine

Looking at what you need to tweak, start again from your success.

What would it take to get you from six days to ten? How about going from six to seven? That's just one more day. After all, this is what you're doing most of the time now, right?

What would it take to get you from 60 minutes to 75? I know there are days where you spend 10–15 minutes choosing your three projects in the morning instead of the night before. That means, on some days, you actually *did* spend 75 minutes, you just used them for something else.

What would it take to be more consistent the night before? You saw what 10 minutes the night before did for your day.

What would it take to reduce interruptions/decisions? Remember the blank page for capturing thought interruptions. I still think this is one of the most valuable tools in my arsenal. If I just capture it, it stops me from opening my email or picking up my phone.

Hint: you might have to rid yourself of the phrase 'this will only take a minute' in your head, or 'I better just do this right now, or I will forget.' You won't if it's written down.

If you've given this a shot and the two hours is working well for you, congratulations!

Identify the next place you can apply *put success in your way* and build your frame for that.

If you haven't tried this, make one of your three things time to measure, dump, and refine your system.

Reflective practice is another way to talk about measure, dump, and refine.

It's less formal than what I've laid out, but it is rooted in the fact that *any* work that we do requires some amount of ongoing reflection to learn from and be more effective.

Dale Carnegie talks about a 'weekly review.' Stephen Covey's habit was 'sharpen the saw.' I could list so many others. Over the next few weeks, look for opportunities for this in your work.

Where can you find time to reflect and learn?

What does that look like?

Schedule 10 minutes into your week. Reflect on a recent experience, a meeting, an interaction, a presentation, or even a workout:

- What aspects went well?
- What were you doing or saying?
- What did you notice specifically?
- Why do you think it went well?

Using the successes, create a simple three-part success framework you could use when you do this again, or with something similar.

In order to have a successful _____, I need to:

1.

2.

3.

You *already* know most of the answers to what makes for a successful experience. You know what went well and you probably know why.

In the same way, you also know what went wrong. We can always come up with a long list. Most items start with phrases like:

- I didn't…
- I should have…
- I can never…

By asking yourself what went well, capturing the answers, and putting them into a framework, you begin to ground yourself in the fact that:

1. You have experienced success.

2. You know what it takes to experience it again.

Take time to listen to yourself.

The right questions are the ones which, when asked, result in an answer that moves you forward in your efforts to be better.

The answers don't always come right away. It's not a perfect formula, *ask this and you will get the exact right answer.*

Asking is only half of the process. We have to listen for the answer.

If you don't, you may not discover the success you are looking for.

If you don't, you may end up repeating the same behavior patterns over again.

You may find success here and there, but the goal is to leverage your success into something that moves you forward.

It will be hard at first. It may feel awkward. And the hardest part is trusting the tiny voice in your head when it responds to the question and gives you an answer worth considering.

But it starts by giving yourself the same attention you would give if a friend were seeking your advice.

After all, you do know what you're talking about.

PART THREE

Systems
that serve

CHAPTER 5

●

Small – Big – Small

6048 miles

The CBS program *Sunday Morning* featured a profile on the journey of Bill Helmreich.

Helmreich is a sociology professor at City College in New York. He wrote a book called, *The New York Nobody Knows*.[18] The book details his journey of walking every block in every borough in New York City. A grand total of 6048 miles.

As if writing a book wasn't a daunting enough task, that's a lot miles to walk for research.

The idea came from a game he used to play with his father, called 'Last Stop.' They would ride the subway to the last stop, get out and walk around. The next time it would be the next to last stop, and so on. This was how he learned about the city in which he lived.

---

[18] Helmreich, W. (2013) *The New York Nobody Knows: Walking 6,000 Miles in the City*, Princeton, NJ: Princeton University Press

What does 6048 miles look like?

It sure sounds like a lot and it is. It took Professor Helmreich four years to complete his journey. He says he walked about 30 miles a week. That's around 120 miles a month or 1500 miles each year. According to my math, that breaks down to about 4.28 miles a day.

Let's break it down a little further.

Given that the average person walks about three miles per hour, that means Professor Helmreich walked for about an hour and a half a day, give or take.

6048 miles is an impressive feat. I don't want to diminish his accomplishment but the concept of committing an hour and a half a day to a goal is perfectly reasonable.

Here's the tease. The show opened by highlighting a story about a man who walked 6048 miles: the entirety of New York City.

Why? Because that's a lot more compelling than, 'man walks about an hour and a half every day and then writes a book about it.' But that's essentially what happened.

What does an hour and a half look like?

For Bill Helmreich, it looked like walking.

His 90-minute walks became an adventure. Each block tells a different story with new characters. The sights and sounds change. Each neighborhood has a distinct energy.

I don't know if you can envision a goal four years out. We sometimes have a hard time thinking beyond much more than a few weeks.

But whether it's four years or four weeks, there's always something that needs to be done each day. There's something requiring just an hour and a half of your time every day of the week that could likely change your life.

Professor Helmreich published his book. He's also been commissioned to write five more: one book for each of New York City's boroughs. I'm guessing that means some more walking, even if it involves covering the same ground.

What's most impressive about his accomplishment is his daily consistency. It's not easy to keep taking small steps each day towards a larger goal. Especially when it seems so far away.

Doing the same thing for 90 minutes each day may not sound exciting, but having walked 6048 miles sure does. The same could be waiting for you.

What's your destination? And, more importantly, what steps are you taking today to get there?

Start where you are

In an email conversation with a new member of our small business mastermind, she was remarking about all the conversations going on in our Slack group. There were so many prior conversations, and she wasn't sure how she would get up to speed with all the discussion threads.

My advice was simple. *Don't try to catch-up.* Start where you are. I gave her a quick recipe for that:

- Introduce yourself.
- Ask for help with what you need.
- Offer help when you can.
- Keep coming back for more of the above.

And here's what I am sure of:

- People will introduce themselves back to you.
- People will help you with questions you ask.
- People will appreciate the help you offer.
- You will always find an opportunity to learn something new, meet someone new, help someone, or be helped by someone.

That's how it works.

And it doesn't matter what you may or may not have missed out on. You can always step in and move the conversation forward.

The feeling that we're always behind happens to us all.

People apologize to me all the time about where they are in their business.

Often it begins with, 'I know you guys say to do *X*, but I haven't been doing it. I know I should, but I just…'

Others apologize for the state of their website or how inconsistent they are with their writing and on and on.

There is no need to apologize for where you are in your business. And *I* certainly don't need you to.

We all have areas of our business we neglect. We all get overwhelmed with choices. We all feel some sense of not having done enough or that we should have done *X*, *Y*, or *Z* years ago. If only we had.

Of course, we are certain our business would be entirely different if we had chosen differently, laying our heads on bags of money each night instead of department store pillows.

Start where you are, here at this moment. Your business is what it is. And only the decisions in front of you are going to move you forward.

In fact, the more time we spend thinking about what we did or did not do in the last year keeps us from doing what we need to do right now.

My business partner Chris Brogan and I host monthly webinars. In a recent webinar he outlined the 'Five Checkpoints in the Customer Experience.' Most of the people who attended were interested in learning more about the first checkpoint, Awareness.

That's where they are in their business. What didn't matter in any of the questions raised on the webinar was what people hadn't done, up to this point.

It also didn't matter that *other* people on the webinar were interested in the second or third checkpoints. That's just where *they* are.

He spent time responding to what each person needed.

Please don't worry about what you haven't done or what you may have missed. Just start where you are.

So, what do you want to work on? Where are you starting from?

What do I need to do right now?

There are always times when we are a little off our game. This happens every now and then and when it does, we may even know the reasons why. But knowing doesn't magically change things.

The month of June happens to be a busy month for my family. Maybe it is for you as well. The end of the school year brings with it a collection of events, celebrations, and activities that turn our schedule inside out.

But we adjust, and make things work. We give time where it's needed and navigate our way through it until life settles down again. In short, we figure it out.

But where does the time come from?

If we're so busy normally, how do we manage to get it all done when our schedules are thrown out of whack? How do we manage to figure it out every time?

In these moments, urgency prioritizes things *for* us. The 'unexpected' gets our attention.

Sometimes it warrants our attention; other times it is completely undeserving.

It isn't necessarily a bad thing to act out of urgency. As it turns out, we can be quite remarkable under pressure. We find ways to make it all work which afterwards may even surprise us.

But operating from a state of continual urgency can spiral us out of control. We are perpetually caught in a state of reaction. The space between stimulus and response narrows to a point where our attention isn't directed, it's diverted.

What do I need to do right now?

This sentence is one of the tools that saves me time and time again. If a flurry of activity or interruptions throws me off my game, I stop, take a breath, and ask myself this question.

If another project lands on my desk and my mind starts to wander into everything I have to do, I stop, take a breath, and ask myself this question.

When I find myself distracted by worry or anxiety about work or my family, I stop, take a breath, and ask myself this question.

In every case, it grounds me. It helps me consider whether or not the urgency I am feeling is warranted.

It helps me to put things in their proper place.

It gives me perspective.

 **The unexpected *still* benefits from a plan**

When you start your day with a clear plan, written out, it's easier to pick up where you left off when interruptions happen.

When you are clear about what is important to you in your business and personal life, it's easier to view even the most unexpected events in the proper context.

What do you need to do right now?

Your day is your week, is your month, is your year

Let's imagine you set a goal to earn $500,000 in revenue for your business this year.

What often happens in our brains is we imagine what it would mean to have $500,000. We may even picture it as though someone will write us a big check after a year of hard work and at the end we shall have $500,000.

Of course, you know this isn't how this works. Knowing this doesn't always wipe away that fantasy in our head, though.

For some people I've coached, this image clouds their thinking, no matter how smart or experienced the person.

'Your day is your week, is your month, is your year' is a phrase my business partner Chris Brogan and I use all the time. Truth be told, I'm not sure who came up with it. I'll happily give him the credit.

The purpose of the phrase is twofold.

First, it's intended to help you understand that the actions you take today (and the decisions you make) have the potential to impact your entire year.

Second, it's meant to encourage you to look out to your year and define success.

It's a perfect place to ask the question *What does that look like?*

If I reframe it to 'What does success look like one year from now?' or '*If* we were to throw a party celebrating a fantastic year, what *specifically* would we be celebrating?'

Looking out a year from now is a useful exercise to identify the specific results you are hoping to achieve. Then we can begin to deconstruct things and break them down in order to align our actions to the goal.

Let's take the goal of earning $500,000 in one year.

We need to break it down into a monthly goal. It will take earning $41,666 per month to achieve $500,000 in a year.

If we break that down further, we know that we need to earn roughly $9,600 per week to get to $41,666 per month.

We've established a weekly revenue target of approximately $10,000.

Now, we need to identify the actions required each day in order to earn $10,000 *this* week and the next.

If this sounds overly simple, it is. It is also this simple: if you can identify which actions you need to take each day to generate $10,000 each week in revenue *and* you execute them, you will indeed earn $500,000 in a year.

The challenge, of course, is identifying the specific actions.

It's also a challenge to stay committed to executing them day after day, week after week.

Identify the actions

If you are responsible for sales, and your product sells for $1000, you know you need to make 10 sales each week to make $10,000.

Let's say your closing rate is typically 40%. This means you have to ask for the sale from 25 potential buyers each week to get 10 sales. That's five customers a day.

In this scenario, your *day* simply must consist of *being in a position* to ask for the sale from five people on average.

If your goal is $500,000 in revenue this year and you knew that meeting with five people a day would accomplish this, it now becomes a matter of setting up your day with the daily goal in mind.

The specific actions you take each day, aligned with your goal, become your year.

More miles or a marathon

The idea of running a marathon is monumental to most people. However, if your goal is to run a marathon a year from now, there are concrete ways to approach it each day.

Let's also give this goal a few more aspects beyond running, because you understand that nutrition, hydration, and fueling are essential for peak performance.

You're not just talking about getting some exercise and going on a diet here, you are committing to training for what is a significant endurance event.

Let's assume you're not setting out to win the marathon. You are, however, trying to cover 26.2 miles in a reasonable period of time. Also, while you know it will be hard, you want to be able to feel comfortable and confident.

You aren't running. You are training.

After a bit of research, you are able to identify a plan that will prepare you for this race. Your job is to follow through on the plan.

Working backwards, you learn that the last two weeks before the race are tapering weeks, and that you should be hitting your peak in the two weeks prior to that, about one month before your race.

Your plan lays out how many miles you should be putting in each week, although this is a plan that ramps up.

Working backwards you see that each week has a target number of miles and that in each week you have some shorter running days, some core workout days, some longer running days, and some rest days.

A note about rest days: in all things (fitness or work), rest days are important. They are there so that your body (and your mind) has time to recover in order to grow stronger. You will not achieve your fitness goals faster by skipping rest days or easy days. You will actually hinder them. Follow the plan.

With a bit more research, you discover a nutrition plan you want to follow as well as a specific hydration plan.

Running a marathon in one year's time now becomes a matter of knowing what the plan is each day, before the day begins.

A full water bottle beside your bed helps you start your hydration plan.

A simple meal plan (the same breakfast and lunch each day) helps you keep to the nutrition plan.

Your fitness is a matter of knowing how far you will run or how long you will train that day.

Laying out your daily plan (including all of what you need to complete it) the night before is incredibly important and then it's a matter of doing today's work. And each day you set up for the next and complete those activities.

Your day is your week, is your month, is your... marathon.

Step by step

John and Dawn Grossman own a restaurant called the Holyoke Hummus Company in Western Massachusetts.

John is one of the most loving and supportive people I know. He's also one of the most resourceful and hard-working. John's always working to make things happen in his community and for his family.

Nothing happens overnight

A lot of his customers have no idea what he's done to make it possible to open the restaurant. All they know is that a new restaurant is opening and they now have another option for lunch.

I thought I'd give you a glimpse of some of the stages John has been through in the past few years:

1. John likes to cook. He makes his own hummus and falafel and gets rave reviews from family and friends.

2. With encouragement from a friend to sell his food, he's invited to a small community event (a local men's basketball league) to try selling his sandwiches and test the market.

3. Armed with no more than a folding table, a deep-fat fryer, and a few other supplies, John's first outing is a success. He's invited back.

4. A few more successful outings and John decides to step up to a food cart. He attends more community events and adds catering. He hires a local artist to create a logo. The Holyoke Hummus Company is born.

5. John tests again, doing a pop-up restaurant every Thursday using a local kitchen and function space.

6. The cart's functionality starts to limit him and John goes on a search for a food truck.

7. John locates the perfect used food truck, does some refurbishing, and hits the streets, expanding the local event and catering schedule.

8. After a good run with the pop-ups and food truck, John finds a great building, the site of a former restaurant in downtown Holyoke. He decides to open the Holyoke Hummus Cafe.

There are so many things I didn't mention. He put in work testing and refining the food. He added to the menu, hired employees, marketed with Instagram and Facebook, and a whole host of other details.

One thing I *will* share is that after all that, he completed his last day at his full-time job. He had been working for another organization that had *nothing* to do with selling food this whole time.

I listed only eight steps on John's journey. The last one hadn't even opened yet, and he *just* left his job. He also has a wife and three children whom he always makes time for.

Here are a few things John's story makes me think about.

Have patience and build: we're pretty impatient. Perhaps because we work online and don't sell falafel, we think we should be able to go from cooking in our kitchen to a restaurant overnight. John worked on every step, testing, adding, and most certainly failing along the way. He was also smart and didn't feel the pressure to go 'all in' before he was ready, keeping a solid job most of the way.

Focus on what's in front of you: when John was working from the food cart, he certainly had thoughts of getting a food truck or opening a restaurant someday. But more importantly than the aspirations, he still got up each day and prepared his cart. He prepped the food, made the hummus, and showed up at event after event.

He worked on growing his business by serving people. He earned more customers through relationships and delivering on the promise of great food and a great experience.

Know what you're selling and to whom: John's menu is pretty simple. He's added a few things over time, but he is focused on his main products.

He feeds the people around him, wherever he is. He doesn't worry what the people in the next town are eating or what another restaurant is selling. He serves the folks in line to get his food, inviting them back or to stay connected, and making sure to let them know where he'll be next.

Turns out it's mostly going to be at the same place now. His new location is across from City Hall and near some office buildings. He chose the new location well.

Congratulations, John and Dawn. I can't wait to see the *next* step. Oops, there I go getting ahead of myself.

Everything is a decision

This is not an exaggeration.

Look around at your desk. Go ahead. I'll wait.

What do you see?

Let me take a guess.

A pile of papers, a few books, a broken pair of headphones, four pens, unopened mail, an old picture, a half-filled notebook, a few random folders, a magazine?

I haven't even mentioned your computer desktop. Open tabs and the 'Updates ready to install' message that you keep ignoring, documents you haven't saved, and the myriad little red dots?

I'll spare you the bit about all the unread messages on your phone. And the state of your inbox.

Sound familiar? Just me?

Are you feeling anxious yet?

Every single thing I listed (and I know I didn't get all of them) requires some amount of our attention and a decision.

Heck, even ignoring each of those items, day after day, involves some level of mental energy.

Because each day your brain sees it, acknowledges it, and chooses to do something else.

Think about that for a minute. The dish you walked by earlier. The sock on the floor. The magazine you haven't read since you subscribed. All of it tugs at your attention, even for a split second, and forces you to consider whether to address it or leave it.

Hundreds if not thousands of micro-decisions every day.

Do you feel tired yet?

Are you still wondering why you can't find time for your bigger goals?

Small, intentional, and focused actions can lead to big results.

Pieces of a puzzle, constructed over time, create the full picture.

We understand this. After all, you don't wake up one day and run a marathon without having put in weeks, if not months, of training in small, consistent efforts.

There's something in the way, though. Actually, there are a lot of small things in the way that keep us from taking action.

We have to face this.

Small

Each one of those tiny decisions clutters your brain. They sap your energy and keep you from ever getting to the big, important decisions you need to make to grow your business.

I'm not just talking about messy desk stuff. We continue to find ourselves, week after week, considering the same silly stuff. We think we're acknowledging the decision, but we're not.

We subject ourselves to the same thought loops, never really acknowledging that there's a more effective decision to be made.

Until we face this and fix this, we're effectively keeping ourselves from directing our attention towards the bigger, important stuff.

Big

With the clutter of small decisions out of the way, you can see your way clear to find the big, business-changing opportunities. You may have a glimmer of what they are. Maybe you've written down some big goals somewhere at some point.

When's the last time you saw them so clearly that the next steps were almost obvious?

When was the last time you felt confident that the decisions you were making and the actions you were taking were leading you towards the business and life transforming goals you know you're meant to achieve?

Small

With clarity on your big picture, you can begin to frame the most critical actions.

Decisions suddenly don't seem as hard. And the long-term impact you want to have naturally prioritizes your time; it guides and frames each decision and action.

The next step is to follow your plan.

The decisions are made. The actions are clear. Keep it small. Trust that the daily actions you make are serving your bigger goals.

And whenever you get off track, start thinking 'Small – Big – Small.'

The clarity of big thinking

My friend Tom is the head football coach for a small high school in Central Maine.

Tom and his team have won two state championships, appearing in four of the past five years. Their most recent title came after the school moved up to a more competitive division. It's clear something is going right.

With teen athletes, you might think keeping things simple, and narrowing the job to its most basic level, would be the best approach. It makes sense. If a player only has to worry about what they are told to do and everyone follows the script they are given, they will be successful.

Tom's approach turns that idea on its head.

Rather than narrowing their focus on small actions, Tom goes *big*.

He immerses his players in the entirety of his system.

In the early days of practice, his approach is to overwhelm them with information.

Then, bit by bit, he shows them how all the pieces of his system work together.

The results speak for themselves.

By being exposed to the *big* picture, they begin to understand how the system works and how each player's role impacts each play.

Eventually, every player on the team understands exactly how their role and the *small* actions they take contribute to the success of the system.

Tom also doesn't worry as much about how precisely someone executes their role, because he knows they understand the purpose of each play.

Instead, what he looks for is whether they are where they need to be at the right moment.

Since players understand how each play is constructed, the goal they are trying to accomplish, and their role, he's able to trust them.

He allows them to make decisions and simple adjustments because they understand what they are trying to accomplish.

His unique approach also means that *any* player can assume *any* position at *any* time and know what is expected of them.

This is the power of understanding the bigger picture and taking small actions that align with your big-picture goals.

Focusing on your goal may be hurting you

Golf's The Masters Tournament is one of my favorite sporting events. I'm drawn to all playoff sports. I love watching the very best compete at that level. How they handle the pressure is compelling.

I've never subscribed to the notion that you should be laser focused on your goals. Instead, I believe you should focus on the actions necessary to achieve them.

The difference is subtle, but important. I was reminded of this in a 2017 interview with professional golfer, Rory McIlroy.

At the time, McIlroy was ranked #3 in the world among professional golfers. At age 28, he had already won three of golf's four major championships. The exception is the Masters Tournament.[19]

After a disappointing tenth-place finish, he was candid about his failure:

> I've been in position before and I haven't got the job done when I needed to... I don't think that's anything to do with my game. I think that's more me mentally, and I'm trying to deal with the pressure of it and the

---

[19] At the time of writing in 2019, he still had not won The Masters Tournament

thrill of the achievement if it were to happen. I think that's the thing that's really holding me back.[20]

At least two elements in that reply are critical:

'I don't think it has anything to do with my game.'

'The thrill of the achievement [is] holding me back.'

McIlroy is known for his preparation, routines, and discipline. He works hard every day. His game is sound. He's confident about that.

Even so, in the course of the tournament he became distracted.

Specifically by *imagining the thrill of winning* the Masters. He switched his focus to the goal instead of the actions required to achieve it.

What's the most important thing for me to focus on right now?

I ask myself this question often. It's part of how I refocus my attention back to the actions.

I suffer the same distractions. I'm a master at projections. I can whip up an Excel spreadsheet and lose myself for days forecasting sales and revenue projections.

---

[20] Rory McIlroy cited in https://inews.co.uk/sport/golf/rory-mcilroy-plotting-low-key-route-masters-redemption-527050

The problem, of course, is it's not real. But our brains (mine anyway) can play some wonderful tricks.

I can get sucked in and imagine what my business or my life will be like once we achieve those goals.

In that moment, the thrill has my attention. I'm not focused on taking the actions necessary get there, making the goal harder to reach.

At the Masters Tournament, there are 72 holes of golf played over the course of four days. To achieve par, you would need to hit 288 shots. The winner, Danny Willett, did it in 283. Rory McIlroy had 289. A mere six shots separated them.

Despite Rory's commitment, discipline, and months of daily training. Despite the soundness of his game. Despite the tens of thousands of shots he had taken leading up to the event, every shot in front of him during the tournament mattered.

I don't know exactly where Rory had his mental slip. I don't know on which shots the thrill captured his imagination and he lost his focus.

What I *do* know is that you and I can learn from this.

We have goals. Some of them are simple and some are likely life changing. But there is always a path. There are always steps we have to take long before we get there. And regardless of how hard we've been working, how close we get, no matter how clearly we can see success just ahead, it's still the steps we need to focus on, not the goal.

Mounting evidence suggests that:

- In the face of too many decisions, you will choose poorly.
- In the face of too many things to do, you will get overwhelmed and perform poorly.
- In the face of too many options, you will weigh out hypothetical trade-offs that can at the least result in dissatisfaction after the fact and, even worse, depression.

My degree is in the field of Human Development. My professional background and experience was in the field of Early Childhood Education for 20 years.

One of the most valuable things I learned from the children I worked with is figuring out how to break things down into small chunks to help accomplish a larger task.

I used this with my own children and it continues to help me get through tasks that I may not enjoy, but have to be done.

The most common example of this came when asking my children to clean their room when they were younger.

They're better at it now, but even my teens still need help laying out a path to the goal every now and again.

Here's an example: we take a look at their room, surveying all that they need to take care of. You were a kid once. You know what this looks like:

- dirty clothes
- messy bed
- playing cards
- Lego

- clean clothes
- various art supplies from past projects
- back packs
- shoes
- wrappers and so on

Knowing full well that my son will be completely frozen in his tracks when I ask him to just 'clean his room,' I ask him to do three things: 'Please *only* pick up your dirty clothes, playing cards and shoes. That's it. Don't do anything else. When you're done, please come see me.'

Each time he comes to me I continue to break it down to two or three things that he can tackle. Pretty soon, the idea of a clean room is within sight. Sometimes after doing just the first thing, you can actually see a change in his demeanor, because he starts to see it as possible.

How it works at work

An example of this is in my work is the project of reviewing our year-end financials.

I do this monthly, but I always want to do a year-end review before we send things off to the accountants.

For some reason, I dread this process. It all seems so large and time consuming and I can feel the resistance mounting in the face of all that has to be done. It feels like hours of my time and I already know I don't like sitting for hours on long projects.

Instead of staying overwhelmed when faced with a project you have to do but may not be entirely excited about doing, it helps to break it down.

In this situation, the first step of breaking it down meant identifying the tools/items I needed.

First on the list was pulling together all of the monthly profit and loss statements and printing them out. These statements, a highlighter, and a pen were going to be the tools I needed to get this done. A browser window open to our QuickBooks account and a window for our bank account and suddenly everything needed to do the work is all laid out.

Honestly, writing this makes me feel a bit silly, but this step was so important for me to get past feeling overwhelmed at the whole project.

I also didn't work on it right away either. I just pulled this together. That was all I decided I would accomplish on that project that day.

Having all of that laid out ahead of time is an awful lot like putting my running clothes next to the bed. It's another example of how to *put success in your way*.

Next I established some simple rules:

- one month at a time;
- most recent to oldest;
- 40 minutes at a time.

I might even make it one of my three things.

Suddenly the project isn't massive, but is a task that needs to be done for 40 minutes in the morning.

It changes from something I've been procrastinating over to something I am doing. The short time spent setting up everything I needed saved days of delaying. And then, it's done.

I know you have goals and projects. I'm going to bet these include words such as website, social media, book, blog, design, sales, revenue, or maybe weight, fitness, stop, start, or eat. Does this sound familiar?

What would it look like to break these down into smaller chunks and call each of those a success?

Weight loss

Any weight loss or fitness program I have ever encountered recommended increasing water intake. The goal might be eight glasses per day.

What if you broke that down and simply started with a tall glass of water next to your bed that you could drink when you woke up?

It may not be eight, but it's more than you've been doing and it's a small step on the way to getting to eight. If you do that for three weeks, I promise you you'll find other ways to drink more water.

Writing a book

There's power in knowing the number. Just knowing the number seems to make it instantly smaller.

Books are approximately 65,000 words. If you write approximately 2000 words a day, in 32 days, you could have a book written.

So, 32 days of writing to have the equivalent of a book. If you really get rolling, there are days when you may write 3000–4000 words and suddenly, you have written a book in less than 20 days.

Before writing this book, it was an idea, a project. At times I was getting that overwhelmed feeling. I had a collection of scattered notes that, while useful, had very little structure.

It was only after I built the frame that I was able to really put it all together. I couldn't begin to write it. I had to start with small bites such as the structure:

- Foreword
- Eight chapters
- Conclusion

The next step was to frame out the subject of each chapter.

Before I knew it, I've broken an entire book down into manageable pieces.

Don't focus on perfection, focus on action

The great thing about the concept of bite-sized chunks is that it doesn't have to be perfect to put a simple framework in place.

Dealing with setbacks and context shifts

Sometimes it's helpful to remember what life was like before I figured out how to *put success in my way*.

I was thinking about this recently, because I needed to flex my willpower muscle to get a few things done.

Several years ago we welcomed our youngest daughter into our family. She's adopted and was three when she arrived in our home.

Along with all of the transitions that one might imagine for a child entering a new family, new language, and new culture, she was also fully embracing all that being three means. Have you ever spent time with a busy three year old?

It was the first time my three older children had experienced this force of nature. They will tell you, it's hard to get much done.

I can tell you that my 'end of the day' planning ended up happening much later than usual. I didn't get to it until around 9:30 p.m. on my couch after all my kids were in bed. But it happened.

On one particular day, I was looking at the next day's schedule and I could see that Tomorrow Guy should expect something outside his normal routine. A visit from my mother and, later, a mid-morning fifth-grade band concert. With that in mind, not much of what would be considered a normal morning was going to occur.

Regardless, I stacked my two hours of focused time exactly as I normally would:

- I identified the three items that would make for a 'successful' day.
- I assigned an amount of time to devote to each one.

- I wrote them out on my daily sheet and scheduled the time.

Even with all that preparation and for a number of reasons, my morning took a bit of a left turn. Then earlier than expected, my mother arrived. All of this set me back about an hour, with a hard stop 45 minutes later.

Frames hold us up

It's a common challenge. The circumstances change and you are forced to adjust. You have to make decisions about what to work on. How should you spend those 45 minutes? The choices are almost endless, right? They used to be.

If you remember the story I shared about prior work habits, you and I both know what would have happened.

I would have opened my email and pecked away at whatever was in my inbox, using it like a to-do list for someone else's priorities.

Then, beholden to the demands of others and an array of mixed emotions, I would start choosing which one I wanted to respond to first. Forty-five minutes later, I would be caught in a back and forth email exchange with someone as though it were a chat and be torn about leaving to get to my son's band concert.

This day was different. My framework supported me.

Yes, I had to make a decision, but the options were narrowed for me by the list of three things I made the night before.

Yes, I only had 45 minutes, but I was able to use that time well. Because I had decided what I needed to do, I was able to focus on one of my projects for the remaining time.

I had to exert some amount of willpower, but less so because the options were clear. I was able to sit down, pick one of the three, and get to work.

When it was time to go, I left knowing that even with shifting context, various circumstances throwing me off, my frame supported me.

Maybe fifth-grade band concerts aren't your priority, but on this day it was mine.

It doesn't matter what those things are or what they look like. What matters is being able to be present because you've built a framework that allows it to happen.

This is the value of your attention and the power of simple decisions.

It is about establishing what matters and making decisions before you have to.

It's about constructing a framework so you can be productive on *your* terms and available for the things that matter to you.

Productivity traps

Productivity traps are activities that we love to do but while they feel like work or they are related to our work, they are actually productivity traps. We justify them because we can make a case for their value, but they keep us from the work required.

One of my biggest productivity traps is budgeting and forecasting. I can spend hours inside of an Excel spreadsheet creating and tweaking various models and their outcomes.

It's real work, but I can play real tricks on my brain while I'm creating it. Simply put, it's a bit of a fantasy and not the work that makes it real.

See if you can identify a few of your productivity traps. Here are some that coaching clients have shared with me. Maybe they sound familiar:

- planning;
- endless goal setting;
- developing character sheets for their book (but not actually writing the story);
- reading every post or article tweeted from folks in your industry because you want to stay current on trends.

The 'Small – Big – Small' framework: the five Fs

On the way to making any significant change in the way we operate, there is always some clutter that's getting in the way of how we ultimately want things to be.

There are things we can do right away to clear out some of that clutter and get some early wins before we make some of the bigger decisions we might need to make:

- Face it
- Fix it
- Find it
- Frame it
- Follow it

Face it - face your decisions

I know you've heard that the first step in solving a problem is acknowledging you have one. Hokey as it may sound, it's useful.

Here's the problem: we continue to find ourselves, week after week, considering the same stupid stuff.

Why do we always put ourselves in the situation of wondering what we should have for dinner?

Facing it means asking: What am I really trying to accomplish? Am I spending time making decisions that warrant my attention?

Fix it - reduce your decisions

The small, daily decisions that occupy our time and take up our attention and energy cause decision fatigue.

Eliminate or reduce your decisions

In recent conversations with some of my coaching clients, we've worked on the process of eliminating or reducing decisions.

I've narrowed the options this way.

The answer to most 'should I do this...' questions can often be reduced to one of three options:

- Yes
- No
- Not right now

This simple framework alone saves so many decision-making calories.

A 'Yes' answer leads to other questions and steps for how, etc.

A 'No' answer is pretty final.

A 'Not right now' answer also requires some follow up, such as when you will consider the decision again and/or under what circumstances.

A quick note about 'Not right now': I always encourage clients to beware of this answer. Are they sure it's the right answer or are they *putting it off* for the time being? If they truly see it as valuable enough to consider at a later time, it's the right answer.

Automate your decisions

Automating your decisions sounds more complicated than it needs to be. But I do like to imagine creating a simple 'if, then' solution.

It's about establishing rules.

Another way to think about this is to consider a handful of daily decisions you regularly encounter.

If you had to program Artificial Intelligence to make the decision for you, what *rules* would you tell it to apply?

- Should I buy this?
- Should I eat that?
- Should I watch this or go to bed?

These are just some simple examples that we encounter day after day. In some cases, you may have some established rules.

- I never make impulse purchases.
- I only eat what is on my approved meal plan.
- I go to bed at 9:50 p.m. every week night.

These are all examples of automating decisions, which in turn reduces decision fatigue.

Find it – orient your decisions to the bigger picture

Lately though, I've been wondering if a year is enough. Should we be thinking bigger than that?

My friend Becky McCray has an overarching vision for her life. She calls it a well-rounded life.

What I like about Becky's idea is that rather than shooting for an outcome, she is taking the time to define *what she wants life to look like.*

Frame it – use the big to frame the small

What makes going out to a year and beyond useful is when we start to work backwards from that point to define progress towards those outcomes.

This is where the month and the week definitions are useful.

The most important aspect of this process is how it informs our daily actions in pursuit of these outcomes.

When we start with annual outcomes and work backwards, the definitions become very clear. We can use those outcomes as a filter for what we should or should not put our time and decision-making efforts into.

We are guided towards the smaller decisions that need to be made and the actions required to achieve the big outcome.

Here are some helpful questions for framing:

- What am I trying to accomplish?
- What decisions do I need to make or what actions do I need to take to accomplish it?
- If I am unsure about a decision, what do I need to know and understand that will help me make this decision confidently?

Follow it – use the frame, stick to the plan

When you get distracted and fall off your plan, that doesn't mean it was a bad plan. It simply means you didn't *follow* your plan.

I was going to add that note later. I was going to lead with how you should follow the plan you've created to achieve the outcomes you determined.

It needed to be said earlier.

Big – Small – Big – Small

For Becky's well-rounded life, she uses her definition of that to drive her decision making. It informs her business decisions, health and fitness, relationships, anything related to how she spends her time.

Recently, she decided to sell her liquor store.

Here's where *Small – Big – Small* and the *five Fs* come back into play.

The *small* clutter and daily requirements of running the liquor store were taking up a lot of her time.

Her *big* (well-rounded life) provided her with a filter. It helped her realize that operating the store was in the way of her having a well-rounded life.

Deciding to sell the store, she now had to think *small* and build a plan to handle the decisions (and actions) that go into the process of selling a business.

Becky's process follows the five Fs formula:

- *Face it*: She had to face the problem that the store was in the way of the outcome.

- *Fix it*: Initially, she reduced the daily drag by eliminating some decisions and automating others. (She added a new POS system a few years back.) That cleared the path for her to…
- *Find it*: Her *big* decision was to value time over the requirements of running the business. She decided to sell.
- *Frame it*: She used the *big* goal to frame up the small steps of a plan.
- *Follow it*: She followed a step-by-step plan to sell the business.

By thinking bigger than a one year goal, Becky was able to approach what feels like big, overwhelming decisions such as selling a business with clarity and purpose.

In the end, that's the ultimate goal.

I want you to make intentional decisions with clarity and purpose.

Everything feels big

A lot of my coaching clients come to me with big decisions they are working through.

In coaching, my role is never to make a decision for you. It's to put you in a position to make the best decision for yourself.

In nearly every experience I have had, some form of Small – Big – Small has come into play. And always in the early days, everything feels big.

The biggest challenge we face is our ability to focus on what matters most in our lives and our work. Maybe it's a project, writing, playing a board game with your child, an important meeting, dinner with your significant other. Perhaps your focus challenges are longer term, such as remaining committed to a goal.

I could list for you all of the distractions we face every day, but you know them already. I could talk to you about your phone, your television, email, social media, what you've left undone, and what's left to do. All of it distracts us.

The more we want to accomplish, the more we look ahead to what our goals are, the more distracted we get. Sometimes it's not something external that distracts us, but our own thoughts about what might be or what should have been:

- If only you had...
- Why can't I just...
- How will I ever...

Stop!

And there is the first part of this frame.

In the first few paragraphs, did you catch yourself wandering off a bit? Did you imagine, even in some small way, the distractions and the battles you have, particularly with your own thoughts?

I did even as I was writing it.

Stop! is a command, an interdiction that cuts through the noise our brains can make. I use it often. It's part of a framework to help me return to focus.

Remember the dangers of 'Let me just do that now...'

It's a great theory. We use it poorly and it is derailing our efforts.

Something comes across your desk, or an idea comes into your consciousness. Perhaps it's something as innocuous as: 'oh, I have to call my husband about the change in plans this weekend.'

You think to yourself, '*let me just do that now*, before I forget.'

So, you stop whatever you are doing and make the quick call to your husband, believing that you will get right back to what you were doing.

While you're waiting for him to answer, you start skimming your email because we can't wait for 10 seconds while the phone rings without trying to do something else. So, let's multitask and see if anything came in while we were working on our project.

You have your conversation, but instead of going back to your project work, you open an email that caught your eye while you were scanning and start to read it.

Let's look at that more carefully.

We started by working on one single project.

Our brain reminded us that we needed to make a call.

In that moment, we are leaving the project to shift our attention to making the call.

While we are calling, we decide to open our email and scan through to see if anything came in while we were working on the original project.

We've gone from focusing on one important project to three disparate and distinct areas.

Let's go a step further.

On the phone call, we are quite likely to discuss more than one topic.

As we are scanning our email, each new message we see in our inbox is also a separate topic.

By giving in to the impulse of 'let me just do that now before I forget,' we have split our attention, not just in three places, but we have now gone through a series of splits that we must wind our way back from.

Stop!

Did your mind drift off again? Mine did. You've done this. I've done this.

We have to dispense with the notion that we can *effectively* allow ourselves to be interrupted and *successfully* return to the task without expending some amount of effort to refocus.

There is a cost to interruptions and our continual 'let me do that now, before I forget' mantra is hurting us more than it is helping us.

The problem is in the statement.

We are worried we will forget because our minds are scattered to begin with. But by giving in to that belief, we exacerbate the issue and perpetuate the problem.

Knowing this is the first step.

When you pull your attention away, when you allow your attention to be pulled away by a thought, an action, or some other interruption, you lose focus and it takes work to get it back. More to the point, we have no idea what we've lost by breaking our focus.

My challenge to you is to let go of the idea that *let me do this now* actually works.

The three Rs and the blank page

The blank page is how I handle the various ideas that pop into my head as I am trying to focus on a project. It's perfect for those, *I need to remember to..., I have to call..., I have to add this..., That's a great idea for another project, I want to share that with....*

Those are the interruptions that pull us into *let me do that now, before I forget.*

Instead of fighting a thought away, allow it to rise to the surface and capture it.

The three Rs are Recognize, Record, Return.

Start your day or your project work with a blank page and a pen next to you on your desk:

1. As a thought comes into your head like the ones I mentioned above, something with the urgency to act, or something you need to remember, *Recognize* it.

2. *Record* it on the blank page so that you can free your mind from having to remember it later.

3. *Return* to what you were doing.

Simple is powerful.

As I am engaged in a project, writing this for example, I've set aside the time for it. I've committed time in my day *just* for this.

However, if I were to allow another idea or something I need to remember to do or even should have done earlier to interrupt me, I will end up derailing my current effort to remain focused and my time will be wasted.

If, however, I write it down and get back to my work, I get more done.

Recognize, Record, Return.

RIRA

This method is used to handle the worry, chaos, and negative talk that enters my mind.

Maybe it starts as a daydream, or a cycle of worry while you're engaged in a project, out for a run, or in a meeting with someone. Brains are tricky. These thoughts seem to come from anywhere.

In this case, your brain isn't interrupting you with something you *need* or *want* to remember. It's pulling you away because what you're trying to do is hard.

It's trying to distract you with something else, something easier or more fun, and giving you every excuse to quit.

The method I use is *RIRA*:

- Recognize
- Interdict
- Refocus
- Act

There are times when I have used this often. One is during physical exercise.

For me, it is swimming. Swimming is the one exercise during which I have the hardest time staying motivated. About a quarter of the way into my workout, the voice starts and the end is all I want.

My mind fills with all sorts of negative talk and other trickery to make me give up.

It tells me I don't have time for this, or I should be back at work. It tells me I'm not good at it, that I should be better than this by now. It tells me to quit, or hints at how it would be okay if I cut the workout short today.

Recognize

Notice this step is part of both methods. We simply have to recognize our thoughts for what they are. We have to see them as clearly as possible before we can do anything else.

It's incredibly important, because it is in that space where a decision can be made. At first, it's a very small space and it's hard to see, but when you do it's time to…

Interdict

Interdiction is a bit more forceful. It's an interruption of the interruption and a complete halting of the activities, in this case the thought pattern.

It can be especially helpful in the case of negative talk or unhelpful worry. You can choose a word or a phrase that works for you. I chose *Stop!* as I demonstrated above. It jars me. It sets the stage for me to take back control in order to…

Refocus

Sometimes, it's as simple as getting my mind back to the work at hand. Sometimes, I have to ask myself a question. I may have mentioned it before. It's one of my best refocusing tools:

> What's the most important thing for me to be working on right now?

When I have the answer, that's when it's time to...

Act

In the case of swimming, my act is to execute *the set I'm in*. If I am working on a project, I need to get back to the spot I left.

Many times, in the *Refocus–Act* stage, I get really small on my actions.

Just finish this sentence. Just finish this pool length. And I build from there.

This works in relationships as well. My head can wander, I can get distracted in thought and instead of *being* with my family. I may be physically there, but get caught up in thoughts or worries that aren't helpful.

RIRA gets me back in the game quickly.

Interestingly, *Rí Rá* is Gaelic for chaos. Draw your own conclusions.

Both methods serve me well as immediate frameworks for maintaining my focus or helping me to refocus.

The first is gentler. Thoughts and interruptions will happen. Recognizing and recording them in order to return to your work is more fluid.

The second is more direct for a reason. Our minds can wander into unhelpful places and take us off track. And we may need that sharp interdiction to help us return. It's a helpful tool. It may be hard at first, but you'll become better with practice.

What does that look like?

Notice when you get distracted. Notice when your thoughts interrupt you and pull you away from what you initially chose to work on. Recognize what happens when you follow that interruption. Where does it lead you? Can you see the space for making a different choice? Maybe it's the space to jot something down. Maybe it's the space for interdiction. For now, just notice.

Try the blank page method for the next week and see what happens.

Note: you can always revisit those items later in the day. I like to schedule a few times. Keep it short. Set a timer.

What would it look like if you felt confident that every day you were working on the right things?

What would it look like if you could just do the work instead of worrying about what else you should be doing?

This is what we hope to gain from the Small – Big – Small framework.

This is why we look out to identify the results you hope to achieve in a year, to understand what that means for each month, each week, and, most importantly, each day.

We want clarity. We want to know and trust that what we are working on each day aligns with our goals.

CHAPTER 6

●

The value of emotional decisions

My friend recently remarked that she doesn't spend a lot of time thinking about goals. She prefers to live with a mindset of 'being better than she was the day before.'

I can appreciate the sentiment. It's a noble aspiration. And after all, I love to help people do their work better. But, by itself, 'better' isn't easy to measure.

I asked her how she knew when she was getting better. She couldn't tell me. And that's because it requires definition.

Defining better

Not only was her desire to be better difficult to assess, she also hadn't addressed a key question.

Did she want to be a better mother, spouse, friend, or co-worker? Was she trying to be healthier, better at managing her finances, or read more often?

It begs the question, 'Better at what?'

And as simple and obvious as that seems, it's not uncommon for people to respond, 'I don't know. I just feel like I should be doing better' or 'It seems like other people are more...'

Sometimes when it comes to goals for our work and life, we know we want more of something or to do better at something. We may even get close to knowing what. But naming it is critical not just to have something to shoot for, but to understand how far away from 'better' or 'more' we might be.

Look at the data

A few years back I mentioned to my friend, Becky, that I had this gnawing feeling that I wasn't a great Dad for my children.

She immediately asked an important question: 'What makes you think you're not a good Dad?'

Note how she shifted me from feeling to thinking.

It was simple. I felt as though I wasn't spending enough time with them.

Her response was: 'Okay, spend the next two weeks tracking the time you spend with them.'

I did. It turns out that I was spending a lot more time than I believed. At the time my schedule was packed, and I was worried I was sacrificing time with my kids. I wasn't.

And the tracking also got me to notice when, where, and how. I was attending practices, games, and performances. I was eating dinner with them and playing games. I was having conversations in the car. I was reading books and tucking them in at night.

Mind the gap

The process forced me to face a few key things.

I had to define what 'better' was.

I had to assess what I was already doing and how much.

The next step is to figure out how much more it would take for me to be better. Was it five hours a week? Was it more reading time? More conversations?

All too often we stand in the middle of our business looking out and wanting more, believing we should do more, have more, be better.

All of that may be true, but we have to start by defining what more is, or what better looks like before we can even begin building a plan to achieve it.

Rituals

For many people, our early experiences with rituals come from religion. Most definitions of the word ritual are rooted in religious practice.

We'll use them elsewhere, but let's start there.

I grew up in a Catholic family. We attended mass, I was an altar server for a short while, and we received religious education through our parish.

As I grew older, the rituals didn't make sense to me. From where I sat, it appeared as though everyone was merely 'going through the motions.'

It was certainly true for me. I didn't get it. I did not have a strong emotional 'connection' to the prayers, or the sitting, standing, and kneeling. How then could anyone else?

I couldn't imagine they were experiencing anything meaningful when they all just looked like robots, going through the motions.

What's the point?

After many years away from the church, I found myself drawn back to it as an adult. As much as the old rituals were off-putting to me when I was younger, I found myself supported by them when I returned.

Rituals serve many functions, especially when we understand their purpose.

Even when we don't, they can have the effect of pointing us in the right direction until we figure it out.

Rituals are habits with a purpose.

They are meant to set you up, to engage in an activity that furthers your efforts to live the life you intend.

What do you intend?

You'll notice I didn't say engage in an activity that furthers your *goals*. Of course, that's part of it.

But over time, as we integrate our actions into our everyday life, it becomes less about goals and more about living and working in alignment with the life you intend.

I didn't say working towards the life you intend.

This is all part of it.

We are not just striving to accomplish something far off.

The point of reclaiming the power of our attention is to experience our days as we hope them to be.

The point is to live our days *as though*, rather than *if and when*.

Before I asked you to consider what you *really* hope to accomplish. Is that clear?

We are building to something, so let's review a few things:

- I asked you to get clear on what you intend.
- I asked you clear the decks, to eliminate distractions.
- We've talked about frames to help you regain your focus when things go astray as they inevitably do.

- We've also talked about reflection and focusing on the 'bright spots,' with the purpose of building on our strengths rather than correcting our weaknesses and moving from a larger goal into the actions required to achieve it each day.

What's next?

To *put success in your way*, we use a physical reminder, such as placing our shoes next to the bed, to support us in accomplishing the daily actions that lead us to our desired outcome.

However, it takes more than just putting on our shoes to get us out the door.

Having everything ready is one thing, but to be perfectly honest, if we put our shoes on before anything else, we're going to have a tough time getting the rest of our gear on.

In the simplest of terms, there is an order of operations to getting ready.

When we know the order of operations, we can construct a routine and, if the act is truly important, we can make it a ritual.

Routines and rituals

I don't want either of us to get hung up on words. However you would like to talk about them is fine. That said, let me share with you how I view them, differently from each other.

I think of a routine as something that is more physical, a physical act conducted by our body. Brushing our teeth each morning is a routine, for example. Getting dressed in the morning is a routine.

Routines have a purpose and can certainly be useful.

Rituals are solemn. They may have physical elements, but they're also a mental exercise. As such, they provide a platform for engaging in something meaningful, and potentially life changing. They help us tune out the distractions, and tune in to the moment and our purpose for it.

Have you ever watched a professional baseball player or a college softball player approach home plate?

Elite players have rituals.

There are actions or a series of actions they perform as they enter the batter's box and face the pitcher.

It's more than just a physical routine.

It has the purpose of helping them focus on the moment at hand.

The same is true for a basketball player at the foul line. They may bounce the ball three times, spin it in their hands, bounce it two more times, close their eyes, open them, exhale, and shoot.

Interestingly, both examples above occur in relatively small moments of time within the context of what are very long games. They occur individually and call for a different level of engagement and focus.

Tennis players are similar. Games can go on for hours, but there are those few moments of time before each shot.

Serena Williams and many other players use rituals before every serve to prepare their minds as well as their bodies.

Rituals don't guarantee success. That would be too easy and we'd probably all use them if that were the case.

That said, what they do is prepare your mind and your body to be in the best alignment to do what you're about to do. That is their purpose.

Involvement matters

Research conducted by behavioral scientists Michael I. Norton and Francesca Gino has demonstrated the positive impact of rituals.[21] While there is more research to be done, Norton and Gino have shown that when the ritual is performed versus simply being observed it improves the overall experience.

Norton and Gino are continuing their research in hope of building on their early findings, but their early findings coupled with the mounting anecdotal evidence of rituals by

---

[21] Norton, M.I. and Gino, F. (2014) 'Rituals alleviate grieving for loved ones, lovers, and lotteries' in *Journal of Experimental Psychology*, 143 (1), 266–272, American Psychological Association

top performers in sports, theater, music, and other disciplines underscores its value.

And this is, of course, what we are talking about. We are talking about *you* performing rituals to improve your experience and therefore your performance of the actual tasks you've identified will help you attain your goals.

So, what does that look like?

Components of a ritual

Rituals are the specific, repeated, and disciplined way in which you perform an action or prepare for it.

Rituals usually have specific physical components, as well as a mental component.

Rituals serve a purpose.

It also helps, when constructing or performing rituals, to limit the need for something external, such as having to have the perfect piece of equipment in order to perform.

One of the most important components is that the control rests with the person performing the ritual. It relies on their actions alone to get mind and body aligned to the goal.

In the case of a baseball player going up to bat, yes, there are tools and equipment that are needed, but those are already there.

The ritual consists of *how* the player approaches the plate. Where they put their feet, their hands on the bat, or what they do in between pitches to prepare for what is to come.

Rituals in action

In the same way, it's important to note that rituals are *not* the same as the process you use to *put success in your way*. That said, you *can* ritualize the act.

In order to improve my efficiency and my mindset for triathlons, I ritualized the way in which I got on my bike.

Having already done all of the preparation of my equipment, selecting the route, scheduling the time, and ensuring everything is ready for me to have a successful ride, my ritual for beginning my ride was also important and very specific.

Every time I went for a training ride on my bike, I set up the equipment exactly as it would be in a transition area.

With everything set and ready, I approached my bike in my bare feet, put on my helmet and buckled it.

I put on my shoes. I put on my sunglasses. I jog my bike to the end of my driveway, jump on, take a breath, hit the timer on my watch and start.

Listing the physical steps or observing the process, it appears to be a routine. The ritual occurs when I use the routine to shift my mindset. Rituals point us to what is important.

Before her first serve, Serena Williams bounces the ball five times. Before her second serve, she bounces it twice. She does this *every* time.

In both cases all of the prep work has occurred. The equipment is already there.

This is different. This is all about ritualizing the steps to prepare your mind.

Preparation

Pick two daily actions (or more if you're feeling ambitious) that contribute to your overall goals. Use the daily actions you identified to *put success in your way.*

Create a short checklist of simple steps you can perform to get yourself in the right mindset before you do something important.

Here are some suggestions:

- *Before giving a speech.* I close my eyes and ask myself two questions. Who am I speaking to? (This helps me focus on my audience.) What do I hope to convey? (While I definitely know this, it's simply an affirmation in the form of a question. In other words, 'you know what you're doing.')
- *Before a race.* I visualize the course and the steps (swim, transition one and all the steps, bike, transition two and all the steps).
- *Before writing.* I close my eyes and ask myself, 'Who am I writing to?' and 'What am I writing about?'
- *Before eating.* Our family says grace together. We always end with a chorus of, 'Help us to be better people, every day.'

It doesn't have to be something you do every day, as is the case with the speech or the race.

However, it is something you can perform each time you practice or rehearse.

How amazing would it be if your rituals prepared your mind to the point that you felt a sense of mastery even before you performed?

Make a list of *all* the opportunities in your day where you would benefit from entering it prepared (or simply having decided ahead of time what you will be doing).

This is similar to the idea of *putting success in your way*. Look for aspects of your day in which you could *decide before you have to*. You don't need to do them all at once. This isn't something that happens overnight, but make a list where you could *prepare* for your time.

Maybe it's deciding what the *very* first thing you will do at work each day for your first 30 minutes. Maybe it's deciding what you will wear, what you will eat. Maybe it's dinner plans or your workout schedule. Whatever you choose is fine.

Pick a few items from your list (or more if you're eager) and work *being prepared* into your day.

Preparation helps me engage in an activity quickly, feeling confident that the right decision about my time has already been made and I have what I need to get to work.

Rules help me stay focused, by blocking out external interruptions especially the ones that *feel* like work or *are* work, so they're easily justified. I keep telling myself, *don't worry, you'll get to it.*

The blank page helps me with the internal distractions. From the siren call of lapsed memory and *let me just do this now,* to the inner critic telling me I can't get it done or I'm doing the wrong thing.

A personal note: All of this takes time. It's not easy and you will miss days, break your own rules, and the day will get away from you. That's fine. Start over.

My day still has interruptions that unravel my productivity, require different decisions, and prevent me from getting a workout in or a project completed. Life happens.

Over time though, it gets easier. The days when it goes well begin to outweigh the days when it doesn't and even on the days that get thrown, you start to realize it wasn't as far off as it used to be.

Making emotional decisions

It's a strange thing to encourage, but there's a strong case for decisions rooted in emotion, more specifically, in something meaningful.

For so long we were told to remove emotion from our decision making.

We're told to bring to bear our cognitive abilities of reason and logic to achieve success.

I'm certainly not advocating we abandon this approach altogether.

I am saying that we need a massive dose of meaning and emotional connection to help us make better decisions.

Change is hard

The list of failed attempts in your life and in my life is probably quite long.

Do you need a list, or can you come up with that on your own?

If you conjure up those failed efforts, I bet it's pretty easy to see that data, reason, and logic didn't always help us.

We may have been armed with information telling us to do the right thing. We may have tried to put all our willpower to use. And somehow we still missed the mark.

But take a moment to look back on your most significant personal and professional accomplishments. If they're anything like mine, they are littered with emotional decisions and actions.

Your path to success had meaning.

It's not that the process wasn't painful or challenging.

It's not that we didn't thoughtfully consider essential data, or take a reasoned approach.

To be clear, using emotions to make decisions does not mean being rash or impulsive.

Instead, we are harnessing the power of emotions.

They are what drives us to change, and they sustain us through the hardest moments.

When we face a difficult choice, our emotions connect us to the meaning of what we're about to do.

Make it meaningful

Since 2007 I have participated in Chris Brogan's 'My 3 Words' project.

The purpose of choosing three words is to use them as signposts throughout the year to keep you on the path you've set.

The words are not goals, but are meant to align you with the intention of your goals. Mine are meaningful. I try to harness the power of my emotions.

If you are trying to improve my health, one of your words might be Play.

It's a great word. Play is a great reminder.

And if we flesh it out a bit to ground it in emotion and meaning, it becomes even more powerful.

Attaching it to an emotional statement such as 'I want to be healthy and fit enough to play with my grandchildren' gives weight to the word.

Now, it's not just a goal I've set, but a story I am writing. This story has characters, relationships, and connections. It has personal meaning.

Logic and reason can tell us it makes sense to live healthier, but our emotions, conjured by the image of playing with a grandchild, can help us make better decisions along the way.

Last Night Guy

Do you know about Last Night Guy?

He's showed up on several occasions lately. I've become a big fan.

I should say, there's a lot of evidence he's been around. His simple acts on my behalf have changed my days. I'm incredibly grateful.

Maybe you're familiar with this person or at least their type.

Mine happens to be male, handsome, witty, and incredibly smart. But I digress…

This is the person who does all the little things that make your life easier:

- They make the coffee the night before.
- They put gas in the car so you don't have to stop on your morning commute.
- They lay out the clothes you need for the next morning.
- And when you arrive at your desk to start your day, this person sets everything up for you to jump straight into work.

Sounds amazing, right? So helpful.

The only person better than Last Night Guy, is Last *Year* Guy.

Last Year Guy is the type of person who set aside money for me in case of an emergency, saved for the trip I want to take, or for my retirement.

It's *this* stuff that elevates him to legendary status.

When the two of them work together, I feel unstoppable.

Compassion

You'll recognize these actions as a form of *put success in your way*.

When I've made decisions before I have to and everything I need is ready, I eliminate the need to rely on my willpower to accomplish a task.

I eliminate all the distracting decisions that can pull me off course in the moment.

But let's be honest. It's not always that easy.

Sometimes we're tired and don't want to make coffee for Tomorrow Guy. He knows how to do it. The stuff is all there, let him deal with it.

Tonight Guy wants want to watch another episode of *The Marvelous Mrs. Maisel* and go to bed.

But it turns out there are two critical emotions at play here that can make all this much easier.

I mentioned how grateful I am for everything Last Night Guy does for me. Even more so, when he knows I have to get up extra early or I have a busy day ahead of me.

It's his awareness of this fact and his compassion for me that fuel his desire and ability to help make things easier.

Put success in your way isn't a one and done approach. It requires a cycle of consistency. Gratitude and compassion are the keys to that cycle.

Recognizing the flawed nature and limits of cognitive functions such as willpower, researchers have shifted their focus.

Recent studies have looked to emotions as a key and have demonstrated that gratitude and compassion have a significant influence on achieving our goals.

In his book *Emotional Success*, psychologist David DeSteno demonstrates how these two emotions improve our self-control.[22]

Simply put, when we experience gratitude, we're less impulsive and make better decisions.

When we experience compassion, this too keeps our more hasty instincts in check.

---

[22] DeSteno, D. (2018) *Emotional Success: The Power of Gratitude, Compassion, and Pride,* New York: Eamon Dolan Books/Houghton Mifflin Harcourt

What's more, the benefits of these emotions are transferable. We don't need to experience compassion for a specific person to help that person.

The act of cultivating compassion in general improves self-control. The same is true with gratitude.

DeSteno notes that people were three times more likely to 'do the right thing' in the moment after experiencing one of these emotions.

Participants in one such study were inclined to save more than double for retirement after experiencing the emotion of compassion.

It's no wonder *put success in your way* works so well.

In many ways, it is a continuous cycle of gratitude and compassion that is fueling consistent action.

When it feels hard to stick to your plans, or you're beating yourself up about your lack of willpower or determination or grit, a more emotional response is in order.

The power of gratitude in its simplest form is that it immediately shifts our perspective.

We can go from being overwhelmed at *everything we have to do* to being immersed in *everything we have*, very quickly.

Personally, I've never been all that good at everyday gratefulness.

I'm more prone to anxiety, never feeling like I've done enough, and wondering what I have to do next.

This is not a good formula for gratitude.

Two simple measures

One

A while back, I finished *Harry Potter and the Sorcerer's Stone* with my youngest child. We'd read 5 or 10 pages a night until she pulled the covers up, laid her head on the pillow, and her eyes became heavy. It was a good place to end.

Each new night I'd start with 'When we last left Harry…' and on we'd read.

Two

Every weekday I get to be the first one up in my house.

For many years my morning routine was to nudge three of my four kids into semi-consciousness and head down to the kitchen.

Every morning I would make a pot of coffee and cook three ham and egg breakfast sandwiches. I'd wrap them in aluminum foil and place them on the stovetop, keeping them warm until it's time to leave and drive to school.

I'm an idiot

I fret. I worry. I constantly wonder if I'm doing enough or have done enough. Part of this is what comes with being a parent.

Like you, I'm pretty tired at the end of the day. Of course, no young child really wants to go to bed. The negotiations start early with my daughter. I don't always have the patience I'd like, but we eventually find our way to bed, the book, and sleep.

Getting up before the sun rises is also not fun. Mornings aren't always smooth with a house full of teens. The harried nature of kids finding clothes, grabbing books, homework, and getting out the door is, at times, maddening.

This is what obscures the simple perfection of these moments.

I might be thinking about what comes *after* she falls asleep or *after* I drop them off at school. And that's why I'm an idiot.

I *get* to send my oldest three off to school with a hot, made-to-order breakfast in hand. I drive them there, sometimes in silence, but always together.

I *get* to spend time reading with my daughter. We talk about what *has* happened and think about what *might* happen. It's just the two of us, together.

Is this not the stuff of accomplishment? Are these not goals worth celebrating or luxuries worth enjoying?

Sometimes the bigger goals we're shooting for reveal themselves in the small details of our daily life. They show up in areas that we don't expect and in subtle, overlooked ways.

And while I'll never stop worrying, at the very least these are moments to be grateful for, every day.

Put success in your way

So at this point you're probably saying something like, 'OK Rob, I can plan all day long. I can come up with a goal and I can even break it down into small pieces. *My* problem is sticking to the plan.'

Does that sound familiar?

I also struggle with this even after years of using these methods. In the end, consistency is what it's going to take to get from the first few daily actions, or even the first few weeks of daily actions, to your end goal.

When I first decided to take up running, I didn't even have an end goal. It was just a daily goal of running one to two miles.

Just prior to that, I had an experience that underscored my need to change my diet and my fitness. It was a combination of being inspired by others, and the realization that I needed to get off the sidelines and be a part of the action.

A personal note: At the age of 15, on a warm, late summer Friday afternoon, I arrived back at my high school from a pre-season football game. It was quite typical except that I had thrown my first ever touchdown pass in this game. All I could think about was getting home to tell my Dad.

Things changed quickly. I went to pick up the car from my mother where she worked. A co-worker and family friend pulled me aside and told me that my father had a heart attack and was at the hospital. My mother was with him, but had left me the car.

In a bit of a daze, but completely unaware of what to do at 15, I took the car, went home to retrieve my younger sister and headed for the hospital.

It was a 30-minute drive and I remember driving on the interstate, speeding past a state trooper who didn't bother to pull me over.

I don't remember much else from the drive, but simply remember arriving at the Emergency Room, seeing the look of profound sadness on my mother's face, and hearing words that merely confirmed what I knew the moment I saw her, that my father had died.

What followed isn't important here and now, except to say this:

My father was only 37 when he died. *His* father was also 37 when he died (also of a heart attack).

To say that there is a weight that hangs over my own health and well-being is an understatement. Still, even with this, I have not always made the best choices. I have not always eaten well or exercised. I have not behaved like a man who should be grateful for each day.

The experience I had before I began running was not profound in an obvious way. It was subtle. It was simply me standing on the sidelines, watching a small local race and waiting for others to finish. I was standing with two other men, both of them smoking, both of them significantly overweight, both of them eating doughnuts, while everyone else was running. I was not smoking. I was not eating doughnuts. But I was standing right there.

From inspiration to consistency

Inspiration can get you started. It's a lot like willpower, though. We can only use it for so long and then we need other systems to support us.

In the case of my personal inspiration, I started running a week later.

Initially I was carried forward by my own excitement for something new, the idea of change, and that I was somehow on a path.

After a few weeks, there were days when I didn't run.

There were days when I couldn't find my watch, or my phone wasn't charged, or my running clothes were dirty, etc.

My inspiration, my excitement, and my *willpower* weren't sustaining my motivation through all of those distractions.

This is why I put my shoes by the bed to take action consistently.

You'll note that I didn't say *accomplish my goal*. We know that working backwards from your goal, into defined daily actions, ensures that your actions are aligned to your goals. Then, when we do the work each day and complete the daily actions, the goals come.

After years of living with an ugly family health history hanging over my head, and years of being a father where the knowledge of that history loomed even larger, it was the combination of the emotional inspiration and the tools to *put success in my way* that sustained my efforts.

I went from not running to completing my first 5k in under two months. I later ran several more 5k and then 10k races, and tried my hand at triathlons. I've since completed a full marathon and five triathlons, including a half-iron distance (70.3 miles).

In that race, on the back of my racing bib were the initials of my father and those of my father-in-law as well.

Long before I completed *any* of my races, and long before I finished the half-iron triathlon, there were months of swimming, cycling, and running, weeks of completing very specific training plans, and days of getting up before my children were awake and after they were asleep to run, ride, or swim.

It took grounding myself in the emotional motivations and identifying the daily actions.

Each night, I decided what my training plan would entail.

I felt compassion for Tomorrow Guy. It's hard to get up and run in the cold Maine winters so I got everything prepared for him ahead of time, to *put success in his way*.

And after each run, I was grateful for the experience, confident I had done what I needed to do on that day for my long-term goals, and encouraged by my success that day.

What does that look like?

The power of emotional decisions is that it gives us a motivational anchor for our goals. It gives purpose and meaning to the 'Big' in the Small – Big – Small framework.

It's up to us to return to our understanding of the core elements of *put success in your way* to identify and decide ahead of time what we need to fulfill our plan, and the daily actions required to achieve our goals.

What is your emotional motivation?

What do you need to sustain your daily actions and stay in alignment with your goals (time, tools, gear, information, etc.)?

How can you *put success in your way* by deciding ahead of time and preparing what you need?

CHAPTER 7

●

Decide before you have to

I give people who are struggling with focus and time management two simple and specific pieces of advice:

1. Decide ahead of time what you want to work on.

If we leave our decisions to in the moment, we aren't deciding, we are reacting. Setting aside time at the end of each day to decide in advance what your agenda will be for the next day helps you focus on the business priorities you have determined.

2. Set rules and follow them.

I've mentioned rules about not checking email or social media until I've put in at least two hours of work on *my* projects. This means no incoming texts and no incoming calls.

At a recent conference I had several conversations with folks who had no idea why they were there. They hadn't figured out which workshops were available or which they wanted to attend.

What I decided prior to going in is that the workshops would provide enough strategy and how-to to last me a while. What I also knew is what I went there to learn, from whom, and why.

Take confident action

As a coach and advisor, my primary role is to put clients in a position to make confident decisions.

We talk about the daily decisions as well as the larger challenges of personal and business leadership. There are two questions I ask to help frame things:

1. What does that look like?

We've covered this, but it bears revisiting.

Many of the ideas and choices in business start with phrases such as, 'I'm think we should…' or 'What if we did more of this?'

The question 'What does that look like?' takes us to a place of action. It moves us from wondering to envisioning.

As you can imagine, it also raises other questions that help clarify the goal, and understand what we need to know before moving ahead.

That's the purpose of my second question.

2. What information would help you make this decision with confidence?

More specifically, it's helpful to know what would have to happen for you say *yes* and what would make you say *no.*

More than likely, you have developed an intuitive framework for making decisions. The goal of these questions is to help you tap into your own experience and the priorities you (often unknowingly) lean towards.

> Give me a lever long enough and a fulcrum on which to place it, and I shall move the world.[23]

As a child, my imagination always conjured images of the long lever and the size of world. But everything hinges on the fulcrum.

You *can* move the world

It's fair to say that *put success in your way* is an approach designed to increase the likelihood of you accomplishing your goals.

That said, it's more specific than that. *Put success in your way* is an approach to help you take action on the steps required to accomplish your goals.

In simple terms, the concept boils down to a few clear actions:

1. Decide ahead of time what you want to do.

2. Have everything you need ready ahead of time.

For example, if I need to run each morning as part of my marathon training, I am more likely to do that if I decide the night before *and* lay out all of my running clothes so I am ready to go.

In Archimedes' statement, the goal is to *move* the world. He doesn't say how far. He doesn't say to what ends. What he

---

[23] A common translation of a quotation attributed to Archimedes

knows is that the right lever and a properly placed fulcrum will make it move.

The fulcrum point is critical

For a lot of people with fitness goals, the biggest challenge isn't the act of running or exercising. Oftentimes, the biggest obstacle to overcome is the space between where they are and getting out the door or into the gym.

If I get out the door or step inside the gym, the chances are far greater that I will do what I'm there for.

While my larger goal may be to run a half-marathon, that only happens if I run consistently. In order for me to run consistently, I need to get into my running gear, and outside.

If I have decided ahead of time that I want to run the next morning, my fulcrum point is making sure I *have everything ready*. I have to *put success in my way*, and leverage this critical moment to set everything else into motion.

The day before

About an hour before you are ready to end your day, write out, by hand, three to four of the projects or parts of projects that you are going to work on for the first two hours of your day.

Be specific and limit your scope to what you can accomplish in a reasonable time period. Example(s):

- Build landing page for new product launch – 30 minutes.

- Update notes from recent meetings in CRM (Customer Relations Management) software – 20 minutes.

Note: If necessary, gather whatever you might need for those projects. Think 'clothes by the bed.'

The day of

Arrive at work, or your workspace.

Review your list.

Start working and vow to work only on those projects from the minute you arrive, for the next two hours.

Do not answer texts until you have finished your two hours.

Do not answer the phone (borrow the two-call rule mentioned earlier if you have to) until you have finished your two hours.

Do not check email, even from your phone, until you have finished your two hours.

Do not check Facebook, Twitter, or Instagram until you have finished your two hours.

Commit to doing these two exercises every day for the next two weeks.

What you choose to do with the rest of the day is up to you.

Identify the decisions you make

Let's be honest. Your day doesn't vary that much. You probably sleep in the same bed. Your clothes are in the same closet or dresser. You brush your teeth at the same sink every day. You eat in the same kitchen and drive the same car to work. This is our reality.

Yet, somehow we feel the need to get creative. We create decisions in areas where we don't need them:

- Have you ever spent time looking for your keys?
- Have you wondered what you should have to eat for breakfast?
- Have you spent time looking for socks, shoes, pants, not entirely sure what you should wear?

Me too.

Why do we do this? Why do we create decisions in areas of our lives where there is already the basic structure for a simple routine? Why do we make it difficult?

Take 15 minutes or so and write down all of the areas of your life in which you're making unnecessary decisions. Identify areas where you could simplify your everyday routines?

Find the simplest path

There's an important reason for this. It's not because I want you walking around your house examining parts of your life, adding to your list of things to do.

No, it's because I want you to train your mind to look for the simplest path. I want you to see for yourself where you add complexity to situations that don't require it and to recognize the sound of your coaching voice and remember it knows what to do next. *Note that I didn't say it always knows the answer, but it does know what to do next.*

The simplest path still requires work to accomplish your goals, but when you identify it, you've cleared the decks, decisions are eliminated, and it's simply a matter of executing.

Here's an example:

If one of my clients tells me that one of their goals is to generate $500,000 in revenue this year, we're going to get simple real quick.

I ask my favorite question, 'What does that look like?' In this case, we return to the process of breaking down the goal using your day is your week, is your month, is your year.

I recently went through this with a client:

Annual goal: $500,000

Monthly goal: $41,666

Weekly goal: $9,600 (or so we thought)

We did this exercise earlier in Chapter 5. Here's where we get more specific:

- Tell me how many weeks you're taking for vacation this year.
- Tell me how many weeks you'll be out for holidays, for more than one day.

- In your business, are there weeks where you *know* you won't sell anything (Christmas, Thanksgiving, New Year's, other holidays in your country)?
- How many weeks do you *really* have left in the year?

For this client, he ended up with about 42 weeks where he felt he would actively be selling.

This changes the number dramatically.

If he had broken his goal down using 52 weeks in the year, he would have ended up with a number close to $9,600/week when in actuality he needs to be shooting for about $12,000 each week.

Depending on the price of your product, that's an enormous difference.

From there, we identified how many calls he needed to make in order to make the 10 sales it would take to reach $12,000.

Because his typical closing rate is around 25%, it turned out to be around 40 calls each week to achieve 10 sales.

That means he has to make eight calls per day.

Is it that simple?

Yes, it's that simple. For this client and his goal of generating $500,000 in sales revenue this year, he *simply* has to make eight calls a day. That's it.

Note: By using a method exactly like this, Franco (remember him?) had his best year ever.

Is it that easy?

Not exactly, because it takes work. But we identified the simplest path (for this client) to $500,000. We used very specific information about his historical closing rates and the price of his products.

He also knew that *making calls = making sales*. Your path to your customers may not be as clear.

More decisions

Lest you think we missed something, there's another decision that this simple path brings up. Sure he '*simply* has to make eight calls,' but whom should he call?

We used the same process to figure that out.

Assuming he has to prospect for the *right* eight folks, it's now a matter of figuring out where to find them.

For him, the majority of the calls were made to existing clients for renewals or upgrades, which is simply a matter of identifying which customers were due for that.

However, here's the *real* trick for making the calls even more simple.

At the end of each day, he chose the eight people he would call the next day.

He identified the names and numbers of everyone on the list with the basic information he needed right next to him.

Do you think those calls were easier to make?

Of course they were because he decided ahead of time and *put success in his way* by:

- knowing that he had eight calls to make;
- making a list of names, contact information, and a few other details the night before;
- scheduling the time;
- making the calls.

When naval officers tell their personnel to clear the decks and prepare for battle, they remove the items they don't need on deck. They don't clear it of everything, the guns are still there.

They have cleared away things they don't need, and have laid in front of them the tools required to do the job at hand.

Take action

I would like you to use a similar process for simplifying one of your goals. Look at your goal as a whole and break it down into smaller daily actions using a similar framework to the one above and asking yourself the question 'What does that look like?'

Rather than have you run your goal through the *exact* formula above, I want you to coach yourself there:

1. Pick a goal you have set for yourself.

2. Identify the daily actions you need to take to achieve it. Put them in the simplest of terms, for now.

I am being (and asking you to be) overly simplistic for a reason. I also want you to listen for your coach's voice.

Start with the final goal. Drill down to the simplest daily actions that you need to take to make it a reality.

Do this for a few different goals.

Clear the decks

Everything between you and the most effective daily action is a decision and a distraction.

If you want to drink more water every day, any other beverage in your refrigerator is a distraction.

If you want to eat better, any food in your house that is not on your plan causes you to make a decision and is a distraction.

If you want to be more productive, anything on your desk or desktop that isn't related to your current project is a decision and a distraction.

I took a look at my desk today and I had a few piles. Nothing major, but it's stuff I mean to deal with, either by filing it away or logging some piece of information.

The problem is, every time I look at it, I have to make a decision. It's subtle, but it's there. I have to essentially pretend it's not there, decide to do it another time, and get to work on whatever project I've decided to do. It's not a goal killer exactly, but it's just another example of something between me and my goals.

Clear the decks – put success in your way

You identified the daily actions that you need in place to accomplish some of your goals. The next question is what do you need in front of you, or what do you need to remove in order to *complete* these actions?

For every situation in which you are trying to make an improvement or work towards a goal, identify what's in the way (physically) and/or what you need to have in front of you to make completing an action easier.

For the eight sales calls, this meant having the names, numbers, and notes for eight people on a list the night before so that the calls were easier to make.

If you're trying to exercise more, have all of your gear ready to go.

If you're trying to eat better, prepare the food you need for your meals.

Remember, *clearing the decks* is as much about leaving the essential tools in place as it is removing what isn't necessary.

We're not immune to needing this no matter how smart we think we are.

Reclaim your decisions

This is how we reclaim control over our decisions.

The effects of reflexive decisions and their evil twin, reflexive actions, are always lying in wait.

We've all been in the situation where our phone battery has died, yet we continually, reflexively, check it, forgetting that it isn't working.

We've all picked up our phone to check one thing and found ourselves four apps deep having completely forgotten to look for the one thing.

These decisions drain us. Rather than acting with intention, we are reacting on reflex.

But how do we reclaim this power?

As with most things, the answers are simple, but not easy.

Decide ahead of time

We get to decide how so much of our day will go. Will we fill our first few waking moments pouring through a slew of notifications? Or will we enter the day in the silence of the morning unencumbered by the demands of others?

Set rules

Your life, your rules. Believe it or not, you can establish boundaries for when you are available to your clients. You can set rules for when you will respond to emails. You get to decide.

And if we indeed have a predisposition to a set of 'default responses,' why not rewrite the defaults?

Design your own set of 'if, then' statements that guide and protect you from reflexive decisions that don't serve you or your business.

Set rules that do serve you by deciding ahead of time. Here are a few to consider:

- I never make impulse purchases.
- I only eat what is on my approved meal plan.
- I go to bed at 9:50 p.m. every weeknight.

Go ahead. Make a list of rules for yourself. What would help shape the experience you want to have in your business, with your family?

Reclaim your decisions, one rule at a time.

Being deliberate with our decisions and actions is essential to accomplishing any goals, resolutions, or other changes we may choose to undertake. Unless, of course, you're content waking up every morning and saying 'Well, I guess we'll see how it goes today.'

We do have a tendency though to leave ourselves open to the whims of others when we lack deliberateness in how we spend our time, what we want to accomplish, and what we need to do to get there.

Consciously and intentionally

That is the essential definition of deliberate. The success we are seeking requires us to look for the opportunity to make decisions that are conscious and intentional, in alignment with our mission and our goals. What's the alternative, after all?

You and I know exactly what the alternative is. At some point, you've entered the day with uncertainty, working on whatever gets put on our plate through emails that come in, a phone call, or a random selection from the two dozen sticky notes on your desk. This isn't conscious and intentional; it is mindless and reactionary.

It's not easy

Deliberateness takes a bit more effort. Setting an intention for our actions and making decisions consciously requires a few things. Above all it requires preparation. Not just in the form of getting everything you need ready, although that's a big part, it requires preparing your mind long before you have to act.

For example, when you are trying to be more focused, it helps to understand the forces working against you. A simple first step is to make a note of the things that distract you during the day.

Decision removal

Most of the distractions in our life are caused by 'in the moment' decisions we shouldn't be faced with anyway. We subject ourselves to decisions that undermine our goals.

Another step in acting deliberately is to find ways to remove 'in the moment' decisions.

Do I eat a banana or a bowl of ice cream? It's easier to choose the banana if the ice cream isn't there.

Do I open Facebook or write a blog post? It's easier to write if you set some rules about how you want to use your time. There's also some software that can help keep you off Facebook entirely.

Do I go for a run or keep working at my desk? Scheduling in the time you will run and having everything you need ready to go can help make that decision easier.

Do I make 10 sales calls today or check my inbox to see if anyone sent me something important? Having names and numbers ready and the time to call helps. Also, establish a rule to only check your email when you need something, not when you're looking for something to do.

Prepare to be deliberate

Someone who goes to the gym each day to train seems as though they have this tremendous power to make the right decision in the moment.

The truth is, they didn't decide to exercise five minutes beforehand.

No, the decision to go to the gym was made the night before, or the week before, or the month before. They already knew it was a 'gym day.' Their clothes were ready. The workout was selected.

So, the deliberate action of going to the gym simply became a matter of completing the decision that was already made, showing up to the pre-determined location at the set time.

If you're attempting to make your best decisions in the moment, the odds aren't with you, or anyone for that matter.

The real trick is to determine ahead of time what matters to us at home, what matters to our business, and how we should be spending our time.

When we prepare and decide ahead of time where we want to direct our attention and energy, acting consciously and intentionally in the moment is a much easier choice.

This is how we find the space between.

Tech rules

People love to talk ask about which apps are best, or which software to use. To me, it's never as interesting as the work of helping people gain clarity so they can take action.

That said, I do want share an approach to technology and more importantly, why it works.

I like to ask myself a few questions when it comes to technology.

- Will this help me do more of what I *like* to do and less of what I *don't*?
- Am I *sure* I'm not justifying getting this shiny new toy? Will it help me focus on what's really important?

Technology should be of service to my lifestyle and not take me away from it.

This or that?

Since I loathe the act of making small 'in the moment' decisions, I try very hard to set up my life to avoid them.

When I'm at my best, I decide ahead of time. My choices become so narrow, the only option is to 'do only this thing because it's the most important thing to do.'

The next best version of this might be: 'Do you want to do this or that?'

Even then, I want more than the options.

I want serving suggestions to make a choice easier.

Imagine an assistant coming into your office and saying, 'It's 9 a.m. Would you like to spend the next 40 minutes writing your next blog post or would you like to make your sales calls?' And they go a step further and lay out a path.

If you want to write the post, here is where you left off, and here are a few ideas for the direction you were going.

Or if you want to make the sales calls, here are the names of three people, their phone numbers, and some notes about them.

By serving up clear options, we go from this or that to helping make the choices easier, moving you to take action.

What can I say?

A great feature in Gmail, for example, is so simple but powerful.

If I open an email, I am presented with three distinct suggestions for my reply. *And* they often make sense.

Much of the email you get requires you to consider a response.

Even simple questions can cause us to delay our replies because we know we have to make a decision.

So, rather than responding in the moment, we pause and think about it. We might even mark it as unread so we can come back later to 'deal with it.'

Sound familiar? Yeah, I know.

But the suggestions Google often serves up are usually relevant to the content of the email. By providing a few contextually accurate reply options, it narrows the decision. So, I go from 'what should I say?' to clicking a box that fits and hitting send.

Google's suggestions also get to the point. The responses are concise. I can always add to them, but the suggestions help get my reply started. It gets me through this part of my day faster and easier.

To me, technology is at its best when it anticipates my needs (even just a little). Serving up helpful suggestions makes taking action even easier.

Decision filters

ActionStacks are a great way to 'automate' your decisions. They allow you to turn what used to be a series of decisions and actions into a simple project plan for repeatable tasks that no longer require a whole lot of consideration.

Add it to the Stack

We all have a tendency to push off certain tasks.

Of course, what makes it even more challenging is that these same dreaded tasks tend to pile up. So, rather than a few simple things we could have done fairly easily, it becomes a large pile of procrastination we have to set aside a significant amount of time to address.

But this isn't what I meant when I said, 'Add it to the Stack.'

The tasks

I'll give you three examples of stuff I tend to put off until another time.

Notes

I'm not a great, 'in the moment' note taker. I find it distracting. I prefer to be present.

That said, my best notes are made in the first 5–10 minutes *after* a call or a meeting.

Cue toddler tantrum.

But I don't wanna do that.

I *want* to move on to the next thing. I'm all done with that other thing.

So, notes pile up.

And speaking of toddlers…

Cleaning up

I just want to eat the things. So, when I'm done cooking and my meal is ready, my natural tendency is to say 'I'll clean it up, later.'

But there's something incredibly satisfying about being able to enjoy the meal without all the cleaning up hanging over my head.

Yet so often my brain rationalizes some burning need to get to the next thing instead.

So, dishes pile up.

Updates

This is kind of a catch-all label for anything that requires a bit of adjusting.

Every so often, something needs to be added or adjusted in our CRM. Maybe an extra step added to a process, or one removed if things change.

Again, the best time for me to do that is in the first 5–10 minutes after I've done the work when the details are fresh in my mind. And also, again, I just want to get to the next thing on my schedule.

So, updates pile up.

Context matters

If I push off my notes, it's harder to remember what *really* happened in that meeting and I'm likely to forget key pieces of information.

If I push off the clean-up, it immediately becomes something I have to make time for later.

If I push off my updates, the next time I need the information it won't be there.

In all cases, the tasks I push off turn into an event all of their own.

So, I add it to the Stack.

ActionStacks

ActionStacks are simple, repeatable plans.

Think of an ActionStack as a recipe card. It tells you what you're making, what you'll need, and all the steps in the process.

When I am scheduling a webinar, I have an ActionStack for that.

Every step in the process is listed. It walks me through what I need to do *and* lets me know about how long it will take.

Most recipe cards have two key time measurements: prep time and cook time.

That's never quite enough. If I plan my cooking time based on *just* the amount of time they list, I wouldn't leave myself any room to clean things up.

We need to leave time around the edges. We need time for transitions.

Recipes have it right on one side, time for preparation, but it takes time on the back end, as well.

I really think it's that simple.

All these leftover tasks aren't separate. They are key ingredients and I need to bake them into the process.

At the end of my coaching calls, I intentionally build in another 10 minutes for note taking. The same is true for any meeting.

When I'm cooking, I can't jump into the next thing, I need time to clean up.

And those updates I mentioned, it's just one more step I add to each ActionStack: 'Check for simple adjustments.' I'm guessing this step adds all of one minute, maybe two, but it saves a whole lot of time down the road.

I think most of what we end up pushing off until later is actually the final steps of just about any small project or task.

Fly the airplane

In his book *The Checklist Manifesto*, Atul Gawande shares his work to bring checklists, like those used in aviation, into *his* field of surgery.[24]

Of course you're thinking, 'don't they already use checklists?' The answer is no.

Despite demonstrating the effectiveness in reducing infections, the need for additional surgery, and other pesky complications like death, he continues to encounter resistance. It's mostly from surgeons who claim to just 'know what they are doing.'

---

[24] Gawande, A. (2010) *The Checklist Manifesto*, New York: Metropolitan Books

The resistance

There are two things he shares that struck me. The first is this idea that we don't like checklists. Gawande (2010: 172) points out:

> It somehow feels beneath us to use a checklist, an embarrassment. It runs counter to deeply held beliefs about how the truly great among us – those we aspire to be – handle situations of high stakes and complexity.

This is despite mountains of evidence demonstrating the effectiveness of using them in myriad situations.

Sure we may know how to do something, but having a checklist removes a level of cognitive workload. Checklists increase our success, by reducing mistakes in key areas.

I use ActionStacks to run many aspects of my business.

They are simple, repeatable plans, our version of a checklist.

In fact, as I was setting a recent webinar, my ActionStack helped identify that a key piece of technology was not working, enabling me to switch to a backup system, for which I also have an ActionStack.

Not every task needs to be a creative endeavor or an opportunity to apply our professional skills.

In the end, we have a job to do and more often than not, a checklist gets it done more effectively.

One of the most fascinating details Gawande found in all of the aviation checklists he reviewed was for engine failure in a single-engine Cessna airplane.

There were six key steps involved in restarting the engine. The very first step on the list was 'Fly the Airplane.'

In the midst of what anyone would deem a critical situation, even the most seasoned pilot needs a reminder to 'Fly the Airplane.'

So often we get caught up in what isn't working in our business or our personal lives. We scramble to get things restarted. We focus more on the impending disaster when the first and most basic thing we should do is 'Fly the Airplane.'

Where could a checklist support your work? What would be the first thing on the list?

Systems as a platform for service

ActionStacks have transformed the way I work. Having a simple project plan to walk through for repeatable tasks has become invaluable. Among the many benefits of ActionStacks, my favorites are:

- *Not relying on my memory* – They offload simple processes so that I don't have to remember each step.
- *Improved decision making* – By having each step pre-determined, I focus my decision-making energy on other things.

These mini-systems are meant to help free us up.

All too often people get stuck in a system loop. They use checklists just to get through them, forgetting what they're really there for.

The system is for service

We've all had a service experience in which someone is simply going through a checklist. They're not even looking at us or hearing what we're saying. They have no desire to listen or engage. They just want to check things off.

The most egregious examples of this happen in healthcare.

Doctors and nurses get overwhelmed with administrative procedures. It's been that way for quite a while. The checklists, rather than being used to support an interaction with a patient, often lack common sense.

In some cases, this creates a barrier to real human interaction.

The spaces in between

We love our children's pediatrician. He has as much on his plate as any doctor does. He has certain boxes he needs to check off during a visit. But what he does so well is use the systems as a platform for engagement.

He doesn't spend time behind a screen, asking questions and typing responses. He doesn't presume what's going on with our child. Even though he's heard the same complaint 10 times that day, he doesn't default to the checklist.

The simple repeatable steps he is required to accomplish don't get in the way, they support him. He doesn't allow it to lead the interaction. The framework is there. It allows him to interact and listen for critical information in the spaces in between.

Use your powers for good

Any system you create, method you adopt, or software you employ should facilitate service. It should keep you doing the work that matters and free you from distraction.

I don't use ActionStacks or *put success in my way* simply to be more productive. I use them as a platform for service.

I use ActionStacks in service to clients and customers. I use them in service to my family. I use them in service to my business goals and my personal goals, and to spend time doing what I love with the people I love.

I use them to make simple decisions in a distracted world to give my time and attention to what matters.

Simple decisions, simple actions

Look at your week.

Remember: 'If you do something more than twice, it needs a system.'

Where are there opportunities to automate your process?

What are the steps? Create your ActionStacks.

CHAPTER 8

●

One number

In your business there is a point that increases the likelihood of your customers buying from you.

If you're a restaurant, that point *might* be after they walk through the door. If you're a consultant or a coach, it might be getting someone to book a 'discovery call.' If you are in sales, it might be getting a decision maker on the phone.

Once those actions are set in motion, the likelihood of achieving your larger goals is increased.

The key is to identify *one number* that will facilitate the greatest growth in your business.

Hint: It's typically not a sales or revenue number.

What is your one number?

It's important to know that when we're talking about this number, it's not the same as your primary goal. It's not your revenue goal, for instance.

Your *one number* is any measurable number that drives an intended goal-related outcome and increases the likelihood of the success of that outcome.

Here's an example:

- Your goal is to have 30 people in every workshop session you run.
- Your *one number* is 10 calls a day. You make 10 sales calls to get people in those seats.
- If you're still not to 30, you can increase that number.

Goal vs approach vs effort

The concept of *one number* separates three areas we tend to group together:

- Goal
- Approach
- Effort

Our goal is to climb a mountain.

There are several paths to get there. Some are difficult. Some aren't so bad. Others are pretty much a cake walk. We choose the one that's not so bad. That's our approach.

We prepare for our climb by training for a month based on the approach we identified. This is our effort.

People tend to worry about their goal and not think much about choosing one approach. Then they apply a lot of effort with no particular approach in mind and they find that they can't reach their goal.

The real trick is picking an approach and sticking with it until you know whether or not you've got an approach problem or an effort problem.

Your one number

To do this, you're going to have to clearly list out your goal and your approach. The approach is where we'll find your *one number*.

If your goal is $120,000 a year in revenue, then your approach has to be something that earns you that $10,000 a month. The approach is *not* the $10,000, because that's just the broken-down goal number.

Let's say you sell coaching at $2500 a month. That means you need four active clients in a month to hit your goal. How much *effort* does it take to land a client? How are you getting them now? Somewhere in those answers is how you'll find your *one number*.

If it's face-to-face meetings, then X amount of face-to-face meetings will be your *one number*.

Your turn

What's your goal? What's your approach? What kind of effort will it take you to get there?

Write it out for yourself somewhere:

- Goal
- Approach ← this is your *one number*
- Effort

Troubleshooting your one number

There's a huge likelihood that you won't pick your *one number* correctly right off the bat. Oftentimes, we accidentally pick the goal measurement and mistake it for the approach number. Remember:

- *Goal* is the finish line – Revenue is a goal. Pounds lost or gained is a goal. Book published is a goal.
- *One number* is the way you get there – 50 courses sold is an approach. Two miles walked is an approach. Five hundred words a day is an approach.

It's up to you

This might be something that takes a little work, but you need this early in the process. The secret of priority management is working mostly on those tasks that move the bigger goals forward. It's not about getting stuff done. It's about getting the right stuff done.

Pick your *one number*, even if you change it a week or two later. It's important.

One job

So, let's apply this to marketing.

If you're a small business owner, marketing is *your* responsibility. You may be of a size where you can hire it out or delegate the tasks, but ownership of the plan is on you.

With that, you also know that you are not a full-time marketing person. You have clients to serve.

Marketing is not your job description. So, we have to find a reasonable balance and develop a simple plan you can consistently take action on. Ask yourself:

> How much time is *reasonable* for you to devote to your digital marketing each week?

The only wrong answer is zero.

The right answer will vary.

Of course your expectations for what you will accomplish with your marketing need to be in line with the amount of time you can spend.

But you don't need to post every second of every day to be able to have an impact. The concept behind *one job* is to help you to narrow your focus to the most important next action that someone should take. This goes for the homepage on your website, or your post on Instagram.

Each element should have *one job*. The job should be simple and it shouldn't try to do more than that.

If everything is set to focus on *one job*, you can build an entire connected ecosystem focused on serving people at each point along the way.

For example, the primary job of my homepage is to get people to click on the button to schedule a discovery call.

There are a number of elements, but it's all geared to the discovery call.

The picture of me is relaxed but professional. I want to put a face to the name.

The video is a way to hear some of my thinking.

You may also notice that I mention my Work Like You're on Vacation course here. But *also* note that I don't have a big flashy button.

Honestly, I don't care if you buy my course here. I am using it as social proof, to demonstrate in some way that I have created something of my own, a method of sorts.

One measure

The beauty of *one job* is that it narrows down what you measure for success. If this page is doing its job well, then my measurement will be based on how many people schedule a discovery call.

Now, there are lots of things to measure, and installing Google Analytics is a must, but let's be clear on one thing (heh).

If Google Analytics tells me that I don't have very many visitors to my site, specifically this page, it does not mean that this page is failing at its job.

What it means is I have to move out to other efforts that will do the one job of driving traffic to this page.

One source (at a time) – effort

As you know, you have a ton of different sources for driving traffic to your site. The most often thought of are your social media platforms and search engine optimization. Let's pick one.

Instagram

Instagram is a pretty good example because it's somewhat limited in what it can do in terms of links. There are two primary features:

- Profile
- Posts (for now, let's include stories in the posts category)

Profile

- What is the *one job* of your profile?
- What is the *one measurement* of your profile?

Posts

- What is the *one job* of a specific post?
- What is the *one measurement* of a specific post?

There are, of course, a variety of post types. You could have a picture, a video, a story (picture or video); you can now add polls and all sorts of things.

But isn't it easier to think about one post at a time?

A few things:

- Gathering followers is not the job of your profile.
- Gathering followers is not the job of *every* post.

For example, a post may have the specific job of featuring a video about the newsletter topic you are writing about tomorrow. You may want to point people to your 'link in the profile' where they can sign up:

- The post points to the profile.
- The profile has the link that points to the page.
- The job of the page is to get people to sign up for the newsletter.
- The job of the newsletter welcome email is to set the stage for what's to come, and on and on.

The real beauty of *one number* (and *one job, one measurement, one source*) is the simplicity. It narrows our focus and aligns our efforts, making it easier to know you're on the right path (or not).

Conclusion: what does that look like?

What does this look like for you?

You probably understand at this point that the way in which you apply these concepts will look different. I don't know what it will look like for you to *put success in your way*. I don't know which decisions distract you or which projects you should be working on each day.

What I do know is that you want to be better. I know that you want to give your attention to your friends and family, your passions, and your work.

I know that you want to reclaim the space between stimulus and response, and direct your time, attention, and actions in deliberate ways.

It starts by looking at everything in your way, getting clear on what you want, and making simple decisions to make it happen.

Look for opportunities

I was becoming concerned by the impact the information I was consuming was having on my mood and how it was shaping my thoughts.

My consumption habits with news and information had become, for lack of a better term, passive.

There were three specific behaviors and habits I noticed:

- Whenever I was driving, I would immediately turn on the radio, usually a station with more news and talk than music.
- When waiting for something, within 10 seconds, I would pull out my phone and start scrolling.
- Frequently, I would leave a sports-talk show on in the background as a form of white noise, while working from home.

In each situation, I was letting information wash over me.

I wasn't consuming with any intention other than to fill the silence or pass the time.

I would also justify my behaviors.

I'd tell myself that I liked to stay informed and know what's going on, even going so far as to say it was my responsibility to stay up to date.

I like to stay connected with the world. I like to be responsive and aware of what's going on with my friends and my family.

But none of this was helping me.

It wasn't helping me in my business or my relationships. It wasn't helping me make better decisions or serve clients. It wasn't helping me be a better father or husband.

What it *was* doing, was creating a loop of stimulus and response.

My response was generally agitated, frustrated, and elevated stress.

I also noticed how many conversations began with, 'Did you hear what happened?' or 'Did you hear what X said today?' or 'Can you believe they did that?'

So, not only was the information I passively consumed taking up my time in an immediate sense, it had a multiplying effect.

I carried these thoughts throughout my day, along with whatever mood it conjured up.

I brought these thoughts up in conversation, commiserating and spreading the information, and my mood, around.

Of course, none of it was all that important. Not in an actionable way.

None of what I heard or read in those moments provided me with useful guidance.

So, I turned it off.

Now I drive in silence. I sit without picking up my phone. I don't turn on background noise.

Here's what I noticed.

The information I need to know finds me. I'll also take a few minutes here to check 'the news.' But it's on my terms, and I get in and get out.

The silence is awkward, but on a long drive ideas rush in, and solutions to problems just appear, making me more creative.

The breaks I take away from my work (usually a computer screen) are real breaks. I can feel my brain relax and rest.

Because I'm not subjecting myself to the manufactured stress of an endless stream of information, I can manage actual stress better.

I read more and write more.

I'm present when I am with other people.

I could go on.

Reclaim the space

I don't think we realize the constant loop of consumption and reaction we're in.

If we do, we haven't accepted its real impact on our psyche, let alone our time. It's as though we've accepted it as, 'the way things are.'

Maybe it's the refrains of 'hustle' that make us think it's okay.

Maybe it's the same tired look we see on everyone around us that makes us think 'it's happening to everyone so it must be normal.'

But we need that space.

When we allow for silence, we create a space for ideas to enter.

When we cut away the noise of constant information, our goals stand out more clearly.

When we make time, we gain time.

And it's in that space where the best decisions are made.

Build from success

You have experienced success in your life. And it doesn't matter how big or small. Use that to identify ways in which you *put success in your way.*

Maybe one of these sounds familiar:

- You got a job.
- You bought a house.
- You lost a bunch of weight.
- You built a business.
- You left a job.
- You landed a client.
- You ran a 5k.
- You wrote *every* day for two weeks.

In every single one of those examples, you can identify success and, more importantly, the actions that made it happen.

And here's where learning from failure... well, fails us.

Let's just look at the very last example: writing *every* day for two weeks.

If you're anything like me, your intention would likely have been to write longer than a two-week stretch. That didn't happen. It's okay.

And by stopping after only 14 days, we would probably chalk it up as a failure. Stuff happens. Life happens. Whatever. We didn't continue. What else is new? (So says the voice in my head.)

Rather than focus on what went wrong on day 15, I believe we can learn much more from how you managed to write each day for an entire fortnight.

Any time you can string together a series of consistent actions, you are building a method for personal success, even if you don't make it all the way.

Always with the questions

In each accomplishment, there is a series of actions. There were steps you took to make it happen. What did that look like?

- What did you do?
- What did you set up for yourself to make it work?
- And then what did you do?

Each of those questions assumes you were in control.

'You do' and 'you set up.' That's ownership. Yes, we get to own our successes as well.

And so maybe your answers are something like this:

- I woke up half an hour early.
- I put it on my calendar.
- I sat in a quiet place.
- I had coffee ready.
- I didn't worry about what I was writing. I just wrote.
- I wrote out 10 ideas each afternoon for the next day.
- I always had a pen and notebook handy.

What any list you make will likely contain is vital clues and patterns for the actions that produced your result.

At the very least, they are a simple process for re-starting whatever you might have finished. At their best, they are a map you can use for other goals:

- Decide ahead of time what I'm doing.
- Schedule it.
- Have what I need ready.

Do you see how that starts to turn into a frame for success? It's not everything. It's not failproof.

But the more you do it, the longer you string the actions together, the more you'll have to build on.

The goal of *put success in your way* and any of the frameworks I share is to help you structure the areas of your life in such a way that you are free to do or pursue the things in your life that matter.

If that is more work, then so be it.

If that is to spend more time with your family, great.

If you want to spend more hours reading, I am with you.

Regardless of what you want to do or what you hope to gain, I hope you are discovering just how powerful simple decisions can be.

Systems serve you

I can't say this enough. Not just for the larger goal, but for each moment.

When I *put success in my way*, it is not so that I will one day complete a triathlon.

I *put success in my way* so that I can make simple decisions in the face of tremendous distraction. I do it to direct my attention to the daily actions that get me ready to complete the triathlon.

See the distinction there?

This system supports you now to be able to reclaim your time and attention.

I'm not sure what *your* goals are and, without sounding dismissive, it doesn't matter. What *does* matter is whether or not you are doing the work now to achieve them.

On the road

I am in my favorite coffee shop in Portland, Maine: Bard Coffee. Despite being away from my office and in another environment, my day doesn't look very different.

My daily sheet was filled out the night before. My pen and sheet with my blank page are next to my laptop, I've set a timer for my first three projects and I am an hour into the day.

Other than putting on a pair of headphones to block out some of the distractions of a busy cafe, my work looks pretty much the same. This system is portable.

Why

The same system, plus a few extra minutes preparing my bag to go out, serves a larger purpose.

It isn't just because I was leaving for the day and I need to be productive on the road. The system that really supported me in this moment was my end of the day routine. It was Last Night Guy.

Preparing and having my bag ready enabled me to focus on getting my children off to school. It needed to be ready so that I could be with *them* and not worry about looking all over the house for my stuff.

It allowed me to be open to the late sleepers (ahh, teenagers) and their misplaced homework.

If I'm harried, or if my stuff isn't done, or isn't ready, how in the world can I give *them* attention?

How could I send them off on their day on a good note?

Sometimes the system is simple and, in this case, so is my emotional reason for preparing the night before. But that doesn't make it any less important.

Systems are there to serve us.

Firetrucks

My best friend, Matthew, was a firefighter in New York City. Yes, he was there on 9/11. He's retired now, living a simpler life in Maine.

Matt was an emergency medical technician (EMT). He was superbly suited for this job. He has many unique qualities, but two of them served him very well in his role.

The first is his ability to assess a situation, determine what was needed, and act. You can imagine how valuable and necessary this would be for a firefighter or EMT.

The second (and I believe the reason he was so good at the first) was his ability to set up systems that serve him.

Matthew has always been able to craft his environment to suit his needs, often hacking existing systems that most of us take for granted. He's purchased pieces of foam to reconstruct the driver's seat of his new truck to better suit his needs and comfort, but I digress.

Matthew is also an artist. I've watched him sit at his art table and demonstrate how every tool he needed was within arm's reach. Each item thoughtfully considered and deliberately placed, based on how often he used them, for purpose and efficiency.

It is remarkable the thought he places in crafting the environment necessary to be successful. In his mind, each time he sat at his table, he wanted to do the work of his art. The space, therefore, had to be set up serve this purpose.

There are two things we can be sure of on a firetruck:

- Every piece of equipment has a purpose and a place.
- At the end of the day the truck is reset and ready for the next call.

The truck, its equipment, and, yes, myriad procedures *are* the system.

They are a shining example of *put success in your way*, but not for just the reasons you think.

The 'why' of a firetruck seems pretty clear. It serves the goal of putting out a fire.

But wait, there's more!

Before that, before the fire is dispensed, the purpose of the system isn't to put out a fire.

The way that firetrucks *put success in the way* is that they support firefighters to do the work necessary to *accomplish the goal* of putting the fire out. This is an important distinction.

The purpose of each truck, and the equipment it carries, is distinct.

Certain trucks are set up to deliver water; others are set up to reach high places. This is their purpose. Each truck is a system designed for one part of a larger job. It is focused on supporting the work that needs to be done by firefighters to:

- contain the fire;
- get everyone out safely;
- treat those who are injured along the way; and
- put out the fire.

Having a goal in sight is clearly important. But understanding *what needs to be done* and *setting up the system to do it* is critical, because after all… you must do the work.

Your goal won't mean anything if you don't do the work. The systems, then, are established to help you make the best decisions in the moment and take the most important actions.

If you want to be an author and publish a book, you have to write every day.

The way you *put success in your way*, then, has to be centered first on writing every day, not on publishing the book.

If you want to be healthier, you have to eat food that's better for you.

The way you *put success in your way*, then, first has to be built to support what goes into your shopping cart, because long before it goes into your mouth, you have to purchase the right food.

Wrapping up or just beginning

By now you've come to understand that you are at the beginning of something entirely new.

As you do the work, as you practice, as you measure, dump, and refine, you will find your own systems built on these foundations.

You're also likely to fall off course and get a bit clunky.

Maybe you've dismissed parts of this for whatever reason.

I would encourage you to start again, to accept clunky just like hitting the wrong note when learning an instrument.

And if you have dismissed it, I would challenge you to consider what is getting in your way. The process does get easier. It becomes more natural and part of how you see challenges and opportunities.

This is *your* system.

I can't define exactly what *put success in your way* looks like for you. This is yours to shape. You've done it before, it's all there.

The work is still hard, but the systems you build should support you each day and sustain your work in the long term.

Sometimes, despite our best efforts, our plan slips. We start falling into old habits. We might even find ourselves filled with doubt or shame for failing yet again.

It can snowball. As one failed effort gains momentum, pretty soon everything feels off track again. Trust me when I say, *I know this all too well.*

Coaching yourself back

In the ideal scenario, the framework I've laid out for you will help you to maintain your focus on your goals. It will help you to be more productive and reduce distractions and reclaim your attention.

But let's be honest, nothing is ideal. We forget. We miss a day. We get derailed. But when attempting anything is worth doing, it requires practice.

Sticking to the plan

There are many reasons people fail to complete new year's resolutions. We are filled with the enthusiasm a change on the calendar brings. We tend to act impulsively. We make grand statements about our new life, and grasp at something completely out of reach.

It's a recipe for failure. On day 10 (or 14 or even 20) of the 'big change,' when the idea isn't new and shiny, we tend to lose focus. We slip a little and then we slip some more and then we give up.

In this case, acting impulsively hurts us. We didn't take time to consider the context in which we are making the change.

Additionally, making grand statements doesn't help you remain accountable, it actually works against you. You announce that you are going to run a marathon to your relatives at Thanksgiving. They will react with awe and pride and encouragement. This gives you nearly the same reaction you'd get if you just completed it. It tricks your brain into believing you've already accomplished it.

Grasping at something out of reach isn't wrong, but you probably need a ladder if you ever want to hold it in your hand.

So, how do we recover from *these* types of slips?

Ask – return to reflective practice

Earlier, we talked about focusing on the bright spots. If you're anything like me, you've slipped up fairly early at some point. You've also experienced some not-so-kind self-talk. A series of self-critical exchanges is likely to go on in your head. But the fact of the matter is you made it 10 days. That's 10 days of commitment.

Rather than dissect why you couldn't make it to 11, figure out how you made it to 10 whole days:

- What about the past 10 days went well?
- What did I do before each day and during?
- What did I do to keep on my plan for this long?
- Why do I think it went well?

It's easy to list all the reasons things went wrong. But a simple 5-to-10-minute exercise could give you a sense of why things went well. How did you make it as long as you did? It gives you a foundation for getting back on track.

Reflective practice revisited

One of the habits we are attempting to integrate into our lives is *reflective practice.*

When my coaching clients fall off their plan, it's usually because they're not giving themselves enough time. The two areas neglected most are:

- reflective practice;
- planning – clearing the decks (put success in your way).

Reflective practice sounds like it takes a lot of time. While it *can* be used that way, there are also ways in which we can add it to the flow of our day in easy ways.

Give yourself space and a bit of silence

We cram almost every waking minute of our days with some form of action or distraction.

This is all noise that clutters our mind and while we *think* it relaxes us, the data shows otherwise. It keeps us in a constant state of sorting through external stimuli and taxes our brains.

We need a bit of silence to allow for us to think deeply and allow our thoughts to sort themselves out.

Mini reflective practice exercise: return to RIRA

Recognize moments in your day when you reach for a distraction such as your phone, the radio, or the TV (especially if you're not sitting and watching).

Interdict with a word like *Stop!* to catch yourself.

Refocus by picking a topic or an idea you've been thinking about or wondering about. Ask yourself how you think you've been doing on X goal or Y goal. It almost doesn't matter where you start because your brain will bring you elsewhere. The point is to let yourself think for a bit in silence, without distraction.

Act, or in this case, don't. One of my favorite phrases shared by a friend of mine is 'Don't just do something, stand there.' The point is to *not* act out of the need to be busy.

For many folks I work with, it can be harder than it sounds. I've noticed myself reaching to turn on the radio many times after only a few minutes of silence in the car.

Oh, and in those moments of silence, listen for your success voice, not the critic.

A few months in

Another common situation is after a few months of things going well, things become complicated. You're finding yourself skipping some of the most important parts of the process.

You've followed it through quite regularly. You *get* it, *and* you've built some impressive methods to keep you working. You've worked on the smaller parts of your larger goals. Things are going fairly well. Well enough so that you start cheating a bit and not follow your plan as strictly.

The reasons for this are varied, but here are a few I've observed.

If the goals are fitness, financial, or schedule related, the minute folks see a bit of wiggle room they look for ways to fill the gap.

You're on a strict budget to save for a house and you discover that you're doing really well, thoughts may creep in about getting a new car or buying an impulse item. After all, you've done so well you deserve a treat. You tell yourself you'll get back on track.

The same is true with fitness and health goals. We are tempted to reward ourselves for our progress so far and cheat a bit.

When we find ourselves more in control of our schedules, we start filling them up again. We say yes to things outside our goals.

This is the reality.

The solution is very close to what I just outlined.

Cheating on our goals happens. When we don't give enough time and attention to evaluation (through reflective practice) and planning (putting success in your way), we lose sight of where we are in relation to the goal.

I already outlined a way to revisit reflective practice. So, let's focus on *putting success in your way*.

In 'clear the decks', the focus of the exercises is to *eliminate decisions*. You *put success in your way* by clearing away distractions and figuring out exactly what you need in front of you to accomplish the actions that will bring you to your larger goals.

This takes time, but not a lot.

Each night, as I end my day, I go through a simple review process for what I accomplished. I look through any notes I took throughout the day and move them to their appropriate place. Then I open my calendar, grab a new sheet and lay out my plan for the next day.

All this takes me as little as 10 minutes and rarely more than 20.

You don't need to go through this process in the exact same way. The point is to *take the time* to:

- *Review* – This is also reflective practice.
- *Decide* – Deciding before you have to (in the moment) improves your outcomes.
- *Prepare* – Laying out everything you need to accomplish the project or task reduces distractions and excuses.

What most people discover in this process is that they fell off their plan because they stopped taking the time to prepare. The old habits returned. It happens when you start to gain some control. You think, *I've got this*. That's exactly when you're susceptible and old habits creep in.

Keep your end of the day habit. Give yourself time to review.

Again, one of the first places to look when you've fallen off track isn't the actions or projects or tasks themselves. It's in the time you are giving yourself outside of them to reset. You need quiet moments without interruption. You need time to *reflect* and time each day to *review, decide*, and *prepare*.

Use the power of simple decisions to reclaim your time and attention.

What's next?

I wrote this book for you. I wrote it for people *like* you and me who, in the face of the ever-increasing demands on our attention, are looking to create space between stimulus and response.

What *you* do next can be as big or as small as you'd like. If you're a small business owner, a solopreneur, or a CEO of a large company, what *next* looks like will be different. If you work in an office and face constant interruptions, or work from home and your biggest distraction is your own brain, what *next* looks like has to make sense for you, your needs, and your goals.

Maybe you'll pick one part of your day on which to focus. Maybe you'll create a handful of ActionStacks to make your days more efficient and effective. Perhaps you'll choose one specific goal to *Put Success in Your Way*, or maybe you'll do something helpful for *Tomorrow Guy*.

One of the easiest next steps is to sign up for my weekly newsletter here: robhatch.com/attention-newsletter. Each week I share new ideas, information, and approaches to implementing the ideas and information I've discussed in this book. When you sign-up, I'll send you a free pdf copy of my daily sheet along with a short video describing how I use it to *decide before I have to* and *Put Success in My Way* each day. If your next step is to explore what one-on-one coaching looks like, you can sign up for a free 30-minute coaching call at robhatch.com/coaching. I'd love to talk with you.

If you'd like to hire me to work with your team, or speak at your next event, visit robhatch.com/speaking to learn more.

More than anything, I'd love to hear from you. You can always drop me a line with any questions you might have at rob@ robhatch.com.

Thanks. I look forward to hearing from you.

CPSIA information can be obtained
at www.ICGtesting.com
Printed in the USA
JSHW030714211020
8852JS00003B/3